THE GEOGRAPHIES OF GARBA

In memory of Michael John Davies

The Geographies of Garbage Governance
Interventions, Interactions and Outcomes

ANNA R. DAVIES
Trinity College, University of Dublin, Ireland

Routledge
Taylor & Francis Group

LONDON AND NEW YORK

First published 2008 by Ashgate Publishing

Published 2016 by Routledge
2 Park Square, Milton Park, Abingdon, Oxfordshire OX14 4RN
711 Third Avenue, New York, NY 10017, USA

First issued in paperback 2016

Routledge is an imprint of the Taylor & Francis Group, an informa business

British Library Cataloguing in Publication Data
Davies, Anna R.
  The geographies of garbage governance : interventions,
  interactions and outcomes
  1. Refuse and refuse disposal
  I. Title
  363.7'28

Library of Congress Cataloging-in-Publication Data
Davies, Anna R., 1970-
  The geographies of garbage governance : interventions, interactions, and outcomes / by Anna R. Davies.
     p. cm.
  Includes bibliographical references and index.
  ISBN 978-0-7546-4433-0 (alk. paper)
  1. Refuse and refuse disposal--Government policy. 2. Recycling (Waste, etc.)--Government policy. I. Title.

  HD4482.D38 2007
  363.72'8--dc22

                                                                      2007025287

ISBN 13: 978-1-138-27656-7 (pbk)
ISBN 13: 978-0-7546-4433-0 (hbk)

Contents

List of Figures

List of Tables

Acknowledgements

I would like to acknowledge the support of the Environmental Protection Agency ERTDI Programme, the Royal Irish Academy Third Sector Research Programme and Trinity College for providing me with essential funding to conduct research that has shaped my thinking within the field of waste governance. In particular I would like to thank the Irish Research Council for Humanities and Social Sciences who provided me with an invaluable Research Fellowship 2005-2006 that allowed this book project to grow roots. That Research Fellowship took me to New Zealand and to the Geography Department of Auckland University. The help I received while at Auckland University and throughout my stay in New Zealand made my research endeavours there a pleasure. Most importantly the legitimacy of my research depends on the goodwill and accessibility of people working on the coalface of waste management. I would like to extend my sincere thanks to all those who have given up their time and shared their expertise during conferences, interviews and site visits. During the preparation of this book, many people have been on hand to offer their advice and guidance. I am grateful to Ashgate and especially to Valerie Rose who took on this project and to Neil Jordan who finally cracked the whip. I would like to express my gratitude to colleagues in the Geography Department at Trinity College Dublin who helped in many ways during the books gestation period. Particular thanks go to Sheila McMorrow for her help with illustrations, but also to David Taylor, Pete Coxon, Robin Edwards and Mabel Denniston for their support and good humour which kept me sane during periods of adversity. Susan Owens, Harriet Bulkeley and Frances Fahy all took on the onerous task of reading early extracts of this book. Their comments much improved the final version, although any errors, inaccuracies or omissions of course remain my own. Finally friends and family have been incredibly patient while 'the book' was in preparation and special mention must go to Homer for distracting Dimitris while I spent long evenings working on the manuscript. My thanks go to all for them for their forbearance and for maintaining an interest in my progress even as I immersed myself into the less than glamorous world of waste research.

List of Abbreviations

CBRO	Community Based Recycling/Resource Organisation
CDB	City/County Development Board
EIA	Environmental Impact Assessment
EPA	Environmental Protection Agency
EPM	Environmental Planning Model
EU	European Union
GAIA	Global Anti Incineration Alliance/Global Alliance for Incineration Alternatives
GEA	Galway Environmental Alliance
GSE	Galway Safe Environment
GSWA	Galway Safe Waste Alliance
IBEC	Irish Business and Employers Confederation
ICT	Information and Communication Technologies
IDA	Industrial Development Agency
IDEA	Irish Doctors Environmental Association
IFA	Irish Farmers Association
IIRS	Institute for Industrial Research and Standards
IPC	Integrated Pollution Control
IPPC	Integrated Pollution Prevention Control
ISWM	Integrated Solid Waste Management
MfE	Ministry for the Environment
MSW	Municipal Solid Waste
NGO	Non Governmental Organization
NIMBY	Not in my back yard
PPP	Public Private Partnership
RCRA	Resource Conservation and Recovery Act
REPS	Rural Environmental Partnership Scheme
RMA	Resource Management Act
SCP	Sustainable Cities Programme (of the UN)
SDP	Sustainable Dar es Salaam Programme
SPA	Social Partnership Agreement
TAN	Transnational Advocacy Network
UN	United Nations

Glossary of Foreign Words
(Irish and Māori)

An Bord Pleanála	Irish planning appeals board
Bunreacht nah Éireann	Irish Constitution
Dáil Éireann	Irish House of Representatives
Hapuu	Māori sub-tribe or extended family
Hikoi	Māori term for a journey, protest march or parade
Iwi	Māori tribe
Kai	Māori term for food
Kaitiaki	Māori term for guardian, protector, steward
Kaitiakitanga	Māori term for stewardship of the environment
Mahinga kai	Māori term for food gathering areas
Māori	Indigenous people of New Zealand
Mauri	Māori term for life essence of a living creature or thing
Oireachtas	Irish National Parliament
Pākehā	Māori term for European settlers of New Zealand
Senead Éireann	Irish Senate
Tikanga Māori	Māori term for things Māori
Tangata whenua	Māori term for people of the land
Taoiseach	Irish Prime Minister

INTRODUCTION

Chapter 1

Garbage and Governance:
An Introduction

Garbage, all I've been thinking about all week is garbage. We've got so much of it, you know? I mean, we have to run out of places to put this stuff eventually (Andie MacDowell, *Sex Lies and Videotape* 1989).

Early one morning I watched from my vantage point as a packer truck compacted my peanut butter jars and chicken bones with those of my many, many neighbours. What had been mine was now, unceremoniously, the city's. It was time to come downstairs, to find out what happened next (Royte 2005, 24).

Until recently and despite its familiarity the garbage (rubbish, trash or municipal waste as it is also known) we generate through commercial and household activities has been considered worthy of little attention except as something to be removed from immediate experience as quickly as possible. As noted by Scanlan (2005, 9) 'garbage is everywhere but, curiously, is mostly overlooked in what we take to be valuable from our lived experiences, and crucially, in the ways we organize the world'. From a management perspective such garbage has tended to be conceptualized as a technical issue, a concern mainly for local authorities with a statutory duty to provide waste collection and disposal. That the production and management of garbage might have political or cultural dimensions was barely acknowledged, leading to its characterization as a 'lost continent' for social scientists (Fagan 2004).

Despite the best efforts of a few academics in sociology, politics, economics and geography (see for example Barr 2002; Boyle 2002; Fagan 2004; O'Brien 1999; Thomson 1979) to highlight the significance of contemporary 'rubbish society' for modern social analysis critical examination of the ways in which our garbage is governed remains embryonic. This is surprising for while municipal solid waste (the formal term for garbage) is not the largest waste stream it is the most widespread being produced by literally billions of people on a daily basis. There is diversity, both in terms of spatial reach and material content, in municipal solid waste that means it demands significant financial and logistical resources to control, collect, recycle and arrange final disposal. Given the extent of the resources required for waste management in recent years attention to it has moved beyond the realm of engineered solutions to become a matter for political consideration within municipal government (the sub-national tier of government), nation states and international organizations. At the same time non-state actors are increasingly familiar participants in discussions about the ways in which waste could and should be governed as well as being active waste service providers.

Despite the groundswell of participants becoming involved in municipal solid waste management the amounts being produced continue to rise across the globe. It is estimated that more than two billion tonnes was produced worldwide in 2006 alone (Keynote 2007) and this waste is not static. Municipal solid waste is increasingly fluid, moving both within and between nation states, traversing administrative and political boundaries and encountering differing management conditions. The manifold costs, to the environment and society, of dealing with such mobile mountains of municipal solid waste are such that '[f]rom centuries of obscurity the waste industry [has] found itself at the hub of environmental argument' (Murray 1999, 20) and it was in recognition of these conditions that the seeds of this book were sown.

This volume will confront the processes of translocalization and politicization that have emerged within the arena of municipal waste by adopting a comparative governance perspective that permits consideration of the multitude of actors involved in waste. In particular it examines the socio-political and spatial dimensions of municipal waste management to complement the dominant technical analyses, essentially paying detailed attention to the geographies of waste governance. As a result this volume expands sectoral coverage and sits alongside other studies of environmental governance that have focused mostly on issues such as climate change or specific spheres of governance such as new social movements, but it also progresses analytical intervention within the field of comparative governance.

The remainder of this chapter provides some parameters for municipal solid waste governance and its geographies. First the concept of waste is defined and dissected with attention to the various classificatory mechanisms that have been developed for its conceptualization. In particular the links between these categories and the evolution of waste management discourses are scrutinized. General definitional matters concerning governance, including environmental and waste governance, are then explored. Drawing these two areas of debate together the final section presents an agenda for a geographically sensitive comparative analysis of municipal solid waste governance.

Waste: Definitions, Classifications and Management Discourses

As a precursor to the development and application of a waste governance analysis it is important to define key terms and concepts. Waste itself, for example, is a word that has multiple meanings and applications. In different contexts it can be used as a verb, a noun or an adjective to refer to thoughtless spending or consumption; the failure to take advantage of an opportunity or a place that is uncultivated, uninhabited or devastated; as well as a catch-all term for unwanted or unusable substances and materials. A number of texts have addressed these wider social processes of wasting (see Girling 2005 and Scanlan 2005) but the focus for this book is on waste as unwanted or unusable materials. Such waste emanates from numerous sources from industry and agriculture as well as businesses and households, it can be liquid, solid or gaseous in nature and hazardous or non-hazardous depending on its location and concentration.

Definitional Debates

It is now a commonly quoted truism that what some people consider to be waste materials or substances are considered a source of value by others. These contradictory evaluations are particularly apparent when comparing different time periods through history, diverse places or disparate communities (Scanlan 2005). The subjectivity of delineating waste means that even at a given moment in one location there can be different interpretations of the value of materials or substances. Supporters of a Zero Waste approach, for example, see the disposal of any materials through landfill or incineration as a flagrant misuse of valuable resources while others might see the reclamation of energy from waste through incineration as a useful form of resource recovery, even recycling. Equally a five year old computer within a European academic institution may be considered redundant (i.e. waste) because of its incompatibility with information technology upgrades, but the same computer may be seen as a fully functioning machine for other community sectors or a source of valuable recyclates for less economically developed societies. This last example is important because waste products are often a combination of materials, some of which might be useful and therefore of value and others not. It is estimated that around half of the materials within modern computers are potentially recyclable with the rest either contaminated plastic, coated with chemical flame retardants, or toxic materials such as lead, cadmium or mercury. How products are recycled into valuable commodities, who undertakes these practices and under what conditions, are increasingly important questions that deserve more detailed attention than can be afforded here (but see Adeola 2000 and O'Neill 2000). Nonetheless acknowledging the different definitional considerations and evaluative frameworks is important not only because it reveals significant details about the differences within and between communities, states and societies, but also because it is a precursor to constructing the kinds of mechanisms for dealing with the materials thus defined.

As waste legislation has emerged in many economically developed countries during the 19th and 20th centuries so the need for more precise definitions of waste has increased because of the financial and legal implications such legislation can have for producers and consumers. In addition establishing agreed definitions of waste is vital to the generation of data about waste and for the planning of waste management activities. Following on from this, definitions of waste have been developed by various governmental and non-governmental organizations. For example, the 1975 EC Waste Framework Directive (75/442/EEC 1975) defined waste as any substance or object which is discarded or which will be discarded. This definition has been amended on a number of occasions to finally read 'any substance or object set out in Annex I which the holder discards, or intends to discard, or is required to discard' (Waste Framework Directive 2006/12/EC). Under this definition once a substance or object is defined as waste it remains so until it has been fully recovered or does not pose any potential threat to either human health or the environment. As with many definitions this European Union (EU) statement requires further clarification and the 'holder' is defined as the producer of waste or the person in possession of it. The broad definition is also supplemented by a list of categories defined in Annex I (see Table 1.1). However these categories were interpreted differently across EU member states and changes were made to the Framework Directive to specify more clearly a list of waste belonging to each of the categories provided. By 2000 a European Waste Catalogue list had been developed incorporating more than 650 waste categories and still this list is not considered exhaustive.

Table 1.1 Annex I of Waste Framework Directive 2006/112/EC

	CATEGORIES OF WASTE
Q1	Production or consumption residues not otherwise specified below
Q2	Off specification products
Q3	Products whose date for appropriate use has expired
Q4	Materials spilled, lost or having undergone other mishap, including any materials, equipment, etc., contaminated as a result of the mishap
Q5	Materials contaminated or soiled as a result of planned actions (e.g. residues from cleaning operations, packing, materials, containers, etc.)
Q6	Unusable parts (e.g. reject batteries, exhausted catalysts, etc.)
Q7	Substances which no longer perform satisfactorily (e.g. contaminated acids, contaminated solvents, exhausted tempering salts, etc.)
Q8	Residues of industrial processes (e.g. slags, still bottoms etc.)
Q9	Residues from pollution abatement processes (e.g. scrubber sludges, baghouse dusts, spent fillers, etc.)
Q10	Machining or finishing residues (e.g. lathe turnings, mill scales, etc.)
Q11	Residues from raw materials extraction and processing (e.g. mining residues, oil field slops, etc.)
Q12	Adulterated materials (e.g. oils contaminated with PCBs etc.)
Q13	Any materials, substances or products who use has been banned by law
Q14	Products for which the holder has no further use (e.g. agricultural, household, office, commercial and shop discards, etc.)
Q15	Contaminated materials, substances or products which are not contained in the above categories
Q16	Any materials, substances or products which are not contained within the above mentioned categories

Source: Adapted from European Union 2006, L114/15.

Waste was not only a concern for the EU however and organizations such as Organization for Economic Cooperation and Development (OECD) and United Nations Environment Programme (UNEP) were also involved in delineating waste definitions. The OECD defines waste as materials that are not prime products (i.e. products produced for the market) for which the generator has no further use for own purpose of production, transformation or consumption, and which he discards, or intends or is required to discard. Wastes may be generated during the extraction of raw materials during the processing of raw materials to intermediate and final products, during the consumption of final products, and during any other human activity (OECD/Eurostat 2007, 277).

UNEP, through the Basel Convention on the Control of Transboundary Movements of Hazardous Wastes and Their Disposal (1989), also adopts a similar definition, but defers to the requirements of nation state legislation with the view that 'wastes are substances or objects, which are disposed of or are intended to be disposed of or are required to be disposed of by the provisions of national law' (UNEP 1989, 6). An important difference here is that while the EU definition intends to be absolute, the Basel definition is relative to the vagaries of legal systems within nation states. Another divergence is the choice of the term dispose by UNEP compared to discard, as used by both the OECD and EU, which opens up a new dialogue about the precise meaning of these terms.

Despite the semantic contestations surrounding detailed definitions of waste in legal and academic spheres, waste is commonly accepted by actors and organizations to incorporate 'materials that are residual to the needs of the individual, household or organization at a particular time and thus need to be disposed of' (Boyle 2001, 73). However, as suggested earlier, achieving general agreement on a broad definition is only the first step in establishing systems of waste governance. It is also necessary to develop a detailed understanding of the characteristics of those materials or substances now defined as waste in order to be able to manage them appropriately. As a result there have been numerous efforts to develop appropriate management systems by classifying either the state or source of the waste materials under consideration.

Classification and Composition

As waste can exist in a variety of states from solid, to liquid to gas and can be generated by a variety of processes, from agriculture to industry to household and commercial, a number of classificatory systems have been adopted (see Tammegagi 1999; Williams 2005). Essentially these systems break down waste in different ways, by its level of toxicity (for example hazardous or non-hazardous), by its chemical composition (organic, inorganic or microbiological) or most commonly by the process that generates the waste materials (such as household, municipal, industrial, agricultural, construction and demolition). This ordering of waste materials has been seen as a means of facilitating the governing of waste in particular ways. Such classification systems have become more important as the problems of dealing with growing amounts and increasingly diverse bodies of waste emerge.

Classification and composition analysis is particularly challenging in the realm of municipal solid waste, the waste sector examined in this book, due to its diversity. Municipal solid waste includes materials produced by everyday activities within communities and it is so called because it is waste that is normally dealt with by municipal services. According to UNEP (UNEP/UNSD 2004) municipal solid waste is derived from the activities of residential dwellings; commercial activities such as shops, restaurants and offices; institutions such as universities and government buildings; and municipal services such as street litter bins and park maintenance.[1] Given the different sources and the multiplicity of activities it is unsurprising that municipal waste comprises a mix of different materials from bulky white goods or old furniture to garden or food waste. While municipal waste is not usually the largest net contributor to the waste stream its diversity and direct linkage to people and places can make management decisions extremely complex.

Although, as Tammemagi (1999) notes, garbology (the study of municipal waste) has yet to reach the glamorous heights of oil exploration or rocket science there has been far more attention paid to the composition of municipal waste following the Garbage Project that was launched by the University of Arizona, USA during the 1970s. The project excavated and catalogued landfill material as an archaeological study of municipal waste composition. They found that around half of all municipal waste was paper, around 13 per cent was organic material, 10 per cent plastics, 6 per cent metal, 1 per cent glass and hazardous materials and the remainder construction debris (Tammemagi 1999, 63). This composition has changed over the years and by the twenty-first century the USA Environmental Protection Agency reported that there was less paper and cardboard in municipal waste (38 per cent), but more biodegradable waste (28 per cent) and more glass (5 per cent) (EPA 2001).

The exact composition of municipal solid waste will, of course, depend to a large degree on the size of the community and the activities conducted in that location. According to the OECD (UNEP GRID-Arendal 2004) for most countries it is organic materials and paper that are the main contributors to municipal waste. In developing countries, large cities generate most of the municipal waste and data are rarely available for rural areas although factors like the type of energy source used for cooking and heating and seasonal differences inevitably play a part in the composition of waste. A trend uncovered by OECD and UNEP studies is that as countries get richer, the organic share of municipal waste decreases whereas paper and plastic increases. Unsurprisingly areas which are experiencing population growth tend to have greater amounts of construction materials in the waste stream, while those which are based on tourist activities as a primary industry will experience more food and paper waste than other locations.

The heterogeneity of municipal waste means that it crosses other classificatory boundaries. For example it is often not simply comprised of benign materials such as paper and glass, or biodegradable substances such as food or garden waste, it can also contain remnants of hazardous materials such as oven cleaners, paint thinners and batteries.[2] Over time this heterogeneity has caused significant debate over the

---

1   In some countries, such as the USA, construction and demolition waste is also included in this category (see Tammemagi 1999), but the OECD/Eurostat definition is used for the purposes of this study. Municipality is the term given to a local political unit that has responsibilities for its local environs be it a town, city, township or county.

2   Again it is hard to establish the exact proportion of hazardous materials in municipal waste although most studies in economically developed countries estimate it to be around one per cent of municipal waste (e.g. Burnley et al. 2007).

governing of municipal waste; that is how the management of municipal waste should be conceptualized and implemented.

Management Discourses and Practices

Although this book is focused on contemporary matters of municipal solid waste governance societies have long created rules about the regulation of waste within communities. Ancient civilizations such as the Minoans created a basic system of burying solid wastes and the Romans institutionalized the first known municipal waste collection where householders threw their waste into the streets to be collected by horse and cart and transported to an open pit (Wilson 1977). Attention to the removal of wastes from communities has also been identified in the roots of other societies such as the New Zealand Māori who have long adhered to a notion of kaitiakitanga (resource stewardship) in order to maintain the integrity of environments. In Māori culture, Papatuanuku (the earth) is extremely important and tangata whenua (local people) have a vital role as kaitiaki (guardians) for it. Waste can reduce or destroy the life supporting capacity of soils by damaging the mauri (life essence) of the land and affecting the Taonga (that which is to be prized or treasured) of resources therefore the places where it is disposed of are considered carefully (Barlow 1991).

In all of these historical accounts the main motivation for regulating waste and ensuring its disposal to suitable locations was concern both for the health of communities and the environments that supported those communities and this was often driven by what Girling (2005, 26) terms the 'politics of disgust'. While populations remained relatively small and dispersed suitable locations for dumping were readily accessible, however as populations grew and processes of urbanization accelerated concentrations of waste increased and suitable locations for disposal became harder to identify. This led to increased incidences of contamination and disease and in Europe led to health crises, such as the plague of 1347, that have been directly linked to poor waste practices (Tammemagi 1999). While the English Parliament prohibited dumping in ditches and public waterways in 1388 the absence of alternative facilities meant that waste was still either being dumped in convenient locations or openly burnt well into the 19th century. At this time sub-national levels of government, or municipalities, were becoming more established in industrialized countries and it was to these organizations that individuals increasingly looked to provide a more comprehensive system of waste management. By the turn of the 20th century the municipalities of many economically developed countries had institutionalized a basic system of waste collection and a variety of disposal mechanisms were being utilized. Landfilling was the most popular method of waste disposal, but in some locations piggeries were developed to deal with food waste while rudimentary waste separation and recycling practices were also visible, particularly during periods of economic recession or war-time scarcity (see Gandy 2001). At the same time systematic burning of municipal waste was being developed with the first incinerator, called 'the destructor', being established in 1874 in the UK followed by the first waste-to-energy plant developed in the mid 1890s (Murphy 1993). By the early 20th century the UK had more than 70 incinerators generating electricity and as the 1930s progressed the USA had at least

600 municipal incinerators on stream. Whatever disposal technique was adopted at this time the level of technological sophistication employed was low. Landfills tended to be established without protective liners or sealed caps to contain the rubbish and they were often located in areas of cheap land or where there were convenient depressions or holes in the ground. Likewise incinerators operated with limited control over either the waste mix that entered the facility or the emissions that were produced by the combustion process. In economically developed countries from the 1950s onwards there was growing public concern and mounting scientific evidence that these waste management options were still causing both health hazards and contamination of environments.

In response, albeit over many decades, engineers and scientists became increasingly involved in the design, location, construction and management of waste facilities. Considerations of accessibility and geology were mooted as important criteria for the siting of landfills and techniques for creating impermeable linings, waste compaction and soil coverage were proposed to contain waste. Meanwhile air emissions standards were being negotiated that required incinerators to incorporate pollution filters, such as scrubbers or precipitators, to reduce air pollution. All the time these technical fixes were being developed the volumes and complexity of waste materials continued to increase as did consumer awareness of the impact that waste was having on environments and health. During the 1970s communities across economically developed countries became more active and organized in opposing the construction of waste facilities as unwanted local land uses, but their consumption of, increasingly packaged, goods did not abate. It was during this era that national and supranational governments became involved in consolidated attempts to govern waste more effectively. The USA passed the Resource Conservation and Recovery Act (RCRA) in 1976 and one year earlier the EU had developed its first Waste Framework Directive (75/442/EEC 1975) indicating that waste was now considered a big enough issue to warrant national, even international, attention. At the same time the private sector were also becoming more prominent in debates about waste problems and potential solutions. In essence waste was becoming big business. Stimulated by wider societal concerns about the exploitation of finite materials, practices of resource recovery (either recycling or waste-to-energy from incineration) emerged as cornerstones of new national and supranational strategies for waste management. Active environmentalists meanwhile were pushing for more attention to proactive waste prevention programmes rather than end-of-stream waste management processes. As with definitions of waste there has been much debate about the meaning of waste management terms, such as recycling and recovery, minimization and prevention, and it is useful to reflect briefly on some of these.

Waste prevention has been defined as a technique, process or activity that either avoids or eliminates waste at its source (Crittenden and Kolaczkowski 1995) and it has been primarily applied to producers. It has however also been applied to the consumer end of the product life cycle in relation to purchasing habits. It has been used to refer to activities such as fixing products rather than replacing them, buying fewer products or ensuring that goods purchased are used rather than discarded, therefore preventing materials becoming defined as waste (OECD 1998). Not all of these preventative activities are permanent conditions of course, for example the reuse of products – re-using plastic supermarket bags and glass bottles or re-treading car tyres – is only possible until the bag fails, the glass bottle breaks or the car tyres cannot be safely re-tread.

In contrast, waste management measures such as composting, recycling, energy recovery and landfill are practices which deal with material that has already become defined as waste by the holder, be that the producer or the consumer. Within the EU energy recovery from waste and waste recycling are considered separately, with recycling being defined as 'the reprocessing in a production process of the waste materials for the original purpose or for other purposes including organic recycling but excluding energy recovery' (EC 1994, Art.3(7)) and energy recovery specified as 'the use of combustible (packaging) waste as a means to generate energy through direct incineration with or without other waste but with recovery of the heat' (EC 1994, Art.3(8)). Supporters of incineration contest this split between recovery and recycling with the argument that energy recovery is itself a form of recycling and should be reclassified as such. The separation of recycling and recovery was discussed extensively during the 2006 revision of the Waste Framework Directive, and debates about whether they should continue to be separate are on-going. These struggles over defining waste management practices are closely tied to the emergence in recent decades of more holistic waste management discourses such as the waste management hierarchy and integrated solid waste management.

The waste management hierarchy, as its name suggests, proposes a ranking of waste management activities from most to least desirable in terms of environmental or energy benefits such as conserving resources, minimising air and waste pollution and protecting health and safety. It emerged in the 1970s when environmental organizations first began to criticize the dominance of disposal techniques in the waste management field (Gertsakis and Lewis 2003) and called for a more differentiated system for managing various types of wastes. Although there have been, and are still, deliberations around the ordering of the waste management hierarchy and the location of various processes in that hierarchy (Boyle 2003; Price and Joseph 2000), the generally accepted format places disposal of waste to landfill at the bottom of the hierarchy, followed by energy recovery through incineration (or similar processes), recycling or composting of materials, re-use of materials in their current forms, with prevention of waste as the most desirable outcome at the top of the hierarchy (see Figure 1.1). The prescriptive aim of the hierarchy is to move from the least to the most desirable waste practices. The hierarchy has been adopted, albeit with slight modifications, by governments in most industrialized countries and the underlying concepts are to be found in international conventions and protocols such as the Basel Convention and EU Waste Directives. Yet the simple framework belies the complexities of operationalizing the hierarchy when the difficult task of comparing economic, social and environmental costs and benefits of particular waste management practices is initiated (Boyle 2003). Within the EU the waste management hierarchy is currently only a guiding mechanism rather than a legal instrument and a such it has been used as a justification, in some countries, for slowly climbing up the waste management steps from landfill to energy recovery and recycling rather than paying attention to the more politically challenging issues of demand management (Price and Joseph 2000). As a result some fear that the most likely outcome of adopting the hierarchy framework as a shaping discourse is a perpetuation of end-of-pipe solutions rather than the best practicable environmental option (BPEO). There are others who argue against the hierarchy as a linear progression model when placing waste management within a sustainable development framework on the grounds that there are some cases when disposal (the least favoured option) might be the best solution if the impact of recycling or re-using materials has a high environmental or economic cost (Schall 1993). These actors frequently call for a more integrated approach to waste management.

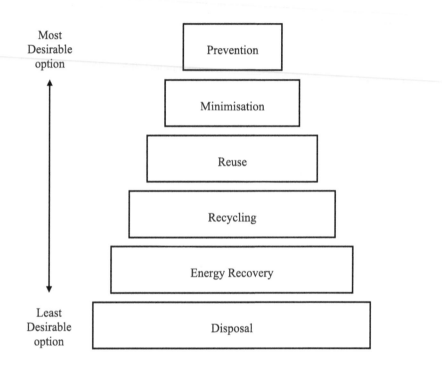

Figure 1.1 Waste management hierarchy
Source: Adapted from Forfás (2003) *Key Waste Management Issues in Ireland*, Forfás, Dublin.

Integrated waste management is the frame of reference that has become adopted across the globe for designing, implementing, analysing and optimising sustainable waste management systems.[3] The approach recognizes the interrelationship of multiple factors in waste management and is fundamentally conceived as the process

---

3   Sustainable itself is, of course, recognised as a hugely contested term, but it has been interpreted in the waste sphere as producing a waste management system that minimises overall environmental impacts including energy consumption, pollution of land, air and water and loss of amenity and operates at acceptable economic cost (White et al. 1995). It should be noted that there is no consideration of social impacts of waste management systems in this definition.

through which both technical and non-technical elements in the management of waste should be considered together to take account of these relationships and interactions (see Figure 1.2). The benefits of integrated waste management are perceived to be manifold with certain problems being more easily resolved when other aspects of waste management are taken into consideration; with capacity and resources being optimized through economies of scale in relation to equipment or infrastructural developments or in terms of balancing costs across the whole waste system and with the participation of actors from public, private and civil society in appropriate roles (UNEP 2005, 8). The means of achieving such integrative waste management are seen as organizational, financial and analytical. Organizational in that there should be some institutional integration of waste-related activities, financial in that costs and revenues from waste activities should be assessed and related, for example using disposal fees to finance recycling or public awareness campaigns, and finally analytical in terms of considering all aspects of the waste system within one planning framework (UNEP 2005). Essentially the language of integration removes the implications of hierarchy in terms of waste treatment and suggests that each management option has a role to play. This pragmatism has ensured that integrated waste management has generated much support from governments and the private sector.

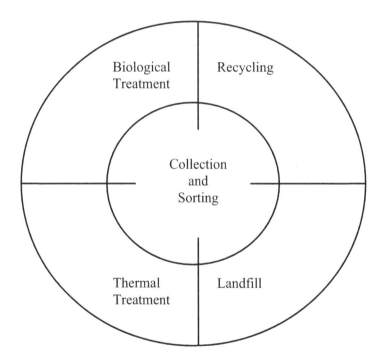

Figure 1.2 Integrated solid waste management
Source: Adapted from ASSURRE, the Association for the Sustainable Use and Recovery of Resources in Europe.

In its ideal form integrated waste management involves a comprehensive life cycle analysis of all products produced. Conducting such analyses is however complicated and the results are frequently contested (McDougall et al. 2001; Seadon 2006). As a result there are those who also see the integration discourse as an excuse to continue with the most environmentally damaging practices of landfilling and incineration while shying away from seriously addressing waste prevention practices, a position expressed by the zero waste movement (Zero Waste New Zealand 2003).

From another perspective there have been calls for all waste to be redefined as resources waiting to be managed. Advocates of this perspective argue that the very terminology of waste management, with all its negative connotations, should be replaced by a broader notion of resource stewardship (see Figure 1.3). Seeing waste as a resource is a pivotal element of zero waste discourses. The zero waste movement, initiated during the 1980s and active in economically developed countries such as Australia, Canada and the USA, identifies waste as an indicator of inefficiency and seeks to alter conceptions with the ultimate aim of eliminating the very idea of waste (Chalfan 2001; Murray 1999). Zero waste statements demand a transition from the current, unsuccessful, position of 'managing' waste to one of eliminating it.

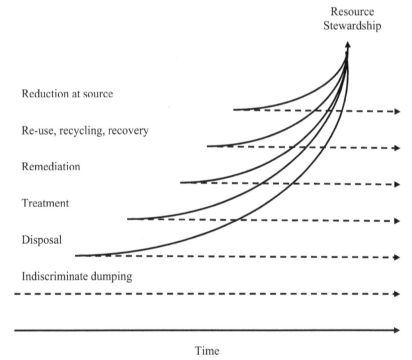

Figure 1.3 Resource stewardship model
Source: Adapted from Stone (2003, 6).

Changing the language used for waste is seen as a crucial, although alone insufficient, step in achieving zero waste such that the movement's documents refer to wasted resources rather than waste and material flows rather than waste streams. As with resource stewardship and integrated waste management the key to achieving zero waste is seen as adopting a 'whole system' approach to redesigning resource flows (Zero Waste New Zealand 2001). The zero waste movement however places a greater emphasis on the need to incorporate social and cultural considerations into that system not just in terms of product design and greater producer responsibility, but also in terms of building social capital as part of developing sustainable local economies through resource recovery systems.

The proliferation of waste management or resource stewardship discourses in recent decades has been spurred on by the lack of success that waste management practices have had in terms of containing waste problems. As mentioned previously the amount of waste and diversity within the waste stream has steadily increased in economically developed countries and many less economically developed countries are also experiencing dramatic rises in the volume and nature of the waste being produced. Although exact figures are difficult to establish OECD surveys have concluded that between 1990 and 2000 EU countries generated 26 per cent more municipal solid waste with only a 3 per cent increase in population, whereas OECD countries exhibited around 20 per cent increase in per capita municipal solid waste generation (OECD 2004). Clearly despite attention to definitions, classifications and management discourses, matters relating to waste and its governance remain problematic.

Governance, Environmental Governance and Waste

The language of governing and governance is explicitly used here to facilitate attention to waste matters beyond local authority disposal mechanisms. These terms permit attention to the various scales, from the local to the global, that may influence the way waste matters are addressed and to the interventions in waste management that emanate from civil society groupings and the private sector as well as public authorities. Essentially a governance perspective allows for consideration not only of technical matters or scientific analyses, but also of the social, cultural, political and economic contexts and networks that shape waste landscapes.

As with the term waste, governance has been defined and applied in a variety of settings at different times generating heated debate about its precise meaning and in particular about the distinctions between governance, politics and the practices of government or governing. Despite this there is common acknowledgement, whether focused on a few individuals, community groups, business organizations, governments or the whole of humanity, that governance analyses are directed towards understanding processes of rule making and decision taking. By logical extension waste governance analyses would seek to understand how decisions are made in relation to waste matters and certainly this is a foundational aim here. However beneath the apparently straightforward task of describing the ways in which decisions are made or practices organized lies a complex of actors and agencies that vie for influence in order to form, inform or reform decisions and decision making environments. Recognition of this complexity has led to the emergence of various governance theories developed with the aim of providing broader conceptual understanding of governance on the one hand and its applications in particular contexts on the other.

Defining Governance

Although a matter of intense and prolonged theoretical and empirical discussion, as Chapter 2 will detail in depth, governance is essentially understood as the manner in which issues (in this case waste) are governed and the respective roles and responsibilities of actors and institutions in practices related to that governing. The etymological roots of the term governance have been linked back to ancient Greece through the works of Plato and his designs for a system of rule through piloting or steering (Kjaer 2004), and elements of this early usage can still be seen in the New Oxford Dictionary of English (ODE) definition that governance is

> the action or manner of governing', where govern is seen as to 'conduct the policy, actions and affairs of (a state, organization or people)' and to 'control, influence or regulate (a person, action or course of events) (ODE 1998, 794).

However since the 1980s political scientists in particular have sought to differentiate governance from the formal structures of government, to admit the participation and influence of non-state actors and institutions from the private sector and civil society.

In the 21st century the term governance has been applied to many different areas and issues. For example, it is increasingly common to come across reference to a normative state of 'good governance' in organizations, such as the World Bank (World Bank 1994) and the EU (CEC 2001), that includes concepts of transparency and accountability in public institutions and decision making processes. For both the World Bank and the EU there are concerns about the effectiveness and efficiency of how decisions are made and executed, who is involved and who is accountable for those decisions and the means by which they are held to account. In essence

> good governance is epitomized by predictable, open and enlightened policy-making, a bureaucracy imbued with a professional ethos acting in furtherance of the public good, the rule of law, transparent processes, and a strong civil society participating in public affairs (World Bank 1994, iv).

While it is hard to argue against these broad statements in isolation, concerns have been expressed about the way in which the World Bank, in this instance, has further interpreted good governance as good neo-liberalism (Wood 2005). As Hirst states 'good governance, therefore means creating free markets, promoting investment, and adopting the right macro-economic policies' (Hirst 2000, 14). In a similar vein the term corporate governance has been developed to deal with issues of accountability and transparency in corporations without fundamentally affecting the practices of business. In contrast governance, as it has been applied to the public sector and particularly in the UK since the early 1980s, has been related to the drastic changes in practices that have come about through the privatization of public owned services or industries and the introduction of commercial management styles into public administration. In other instances governance has been applied to new patterns of interaction, such as networking, partnership or deliberative forums, between actors and organizations in the practices and processes of decision making. Such networks are particularly visible in the field of global environmental problems where it has been recognized that issues such as climate change cannot be addressed by individual nation states. Within this context attention has been increasingly paid to international agencies, regimes and agreements as new forms of environmental governance.

Environmental Governance

It is now common to find analysts of environmental problems engaging with governance as a means of comprehending and evaluating the complex processes of policy making and implementation. Governance has been adopted and adapted in the environmental sphere not only to accommodate non-state actors (Grossman 2000; Keck and Sikkink 1998; Kolk 2000; McCarthy 2005; Newell 2000; Princen and Finger 1994; Roome 1998), but also to engage with issues of scale and place through the concept of multilevel environmental governance (Bulkeley et al. 2003). Although environmental issues are not unique in exhibiting transcalar qualities, the debates that have emerged alongside an increased awareness of environmental problems present a particularly vivid and explicit example of scalar transcendence and multidimensional politics (Auer 2000; Bulkeley and Betsill 2003). Research concerned with issues as diverse as climate change, biodiversity and ozone depletion (Keck and Sikkink 1998; Litfin 1998; Schreurs and Economy 1997; Young 1997) all debate the extent to which the nation state is the primary actor in environmental governance, and the implications, both for practice and analysis, of including non-state actors and sub- or supra- level government in conceptual frameworks of governance. Young (1997), for example, concludes that in order to assess whether international environmental regimes are effective, more attention needs to be paid to the 'bottom up' processes through which regimes are formed and implemented.

Much of the recent work on environmental governance, which is considered in more detail in Chapter 2, presents a challenge to the dominance of international relations theory in examining global environmental politics that has tended to reify the nation state as the sole locus of power in global governance. More nuanced approaches have been proposed that recognize the ways in which international institutions and norms are constructed and contested by multiple actors working at a range of scales (Bulkeley and Betsill 2003; Kutting 2000; Paterson 1996). Yet while there is growing attention to environmental governance, the analysis of these processes and practices have been dominated by a small number of issues that have been the subject of intense global consideration, such as climate change. There has been less attention given to waste governance and particularly the governance of household waste.

Waste Governance

As has been implied by earlier sections of this chapter, the ways in which waste is managed is already a matter for international concern as indicated by the attention of organizations such as the OECD, UNEP and EU to waste issues. Although there are far fewer international regimes in this field compared to climate change, biodiversity or ozone depletion, the growth rates in waste production and the emergence of waste flows around the world – or waste tourism – suggest waste will only continue to grow as issue of global significance (Fagan 2004; Greenpeace 1993; Wapner 1998). There is also greater evidence of national strategies for waste management feeding into and also being affected by the findings of these international discussions. Countries as diverse as South Africa (Department of Environmental Affairs and Tourism 1999) and Scotland (SEPA 2003) have already developed national waste management plans or strategies with others like Chile (CONAMA 2006) following suit. At the same time public-private partnerships for waste service delivery are becoming increasingly common in both more and less economically developed countries (see Ibitayo 2002; Rathi 2006; Rosell 1996). Networks of waste organizations, such as the International Solid Waste Association (ISWA), are attempting to represent the interests of participating institutions in policy debates and transnational advocacy networks, such as the Global Alliance for Incineration Alternatives (GAIA) and the Zero Waste Alliance, seek to support localities in opposition to incineration technologies and in attempts to move towards zero waste approaches (Davies 2005). So while traditionally seen as a matter of disposal for local governments the production, transport and treatment of waste, with its identified impacts on social, political, economic and environmental conditions, appears now to also be a matter for nation states and supra-national entities, the private sector and civil society (Bulkeley et al. 2005). Waste issues are present in discussions involving different spheres of governance (public, private and civil society sectors) at a variety of scales from the local to the global (i.e. across the tiers of governance) and the tiers and spheres of waste governance are considered in a range of different national contexts in Chapter 3.

It is possible then to identify the existence of different tiers and spheres of waste governance, but how do these tiers and spheres interact, how do these patterns of interaction emerged and how will they evolve in the future? To answer these questions an agenda for analysis is required and the next section outlines the approach adopted here to facilitate a consideration of municipal solid waste governance in a comparative context.

Conclusion

To reiterate, the aim of this volume is to provide a critical account of the governance of municipal waste in different locations by examining public sector practices in the waste field, from the local to the supra-national level, and the interactions between these public sector practices and the activities of private and civil society sectors. This interaction between the tiers of government and spheres of governance, frequently termed multilevel governance (Bulkeley et al. 2003), provides the starting point for the analysis. The contested concept of scale is then a core issue and following Whitehead (2007, 27), 'scale is not understood here simply in terms of relative size (something like the region being bigger or smaller than something else – like the city or state), but as a relational category', where relational refers to the interconnections between processes operating globally, nationally and sub-nationally. Matters of scalar interactions sit alongside attention to interactions between society and environmental resources, economy and ecology, politics and nature that lead to the production of waste in the first place (Whitehead et al. 2007). These concerns make the project fundamentally geographical in perspective as the production of waste and its management create a spatial signature on the landscape both physically and conceptually.

In Part 1 attention is given to waste governance theories, concepts and practices in order to set the foundations for an analytically rich and empirically grounded investigation of waste governance. The conceptual frameworks that have developed in the field of governance, environmental governance and the governing of waste are considered in Chapter 2, while Chapter 3 reflects on the study of waste management in different continents and countries. A number of key issues emerge from these studies, not least the need for a research approach that can accommodate processes, practices and outcomes of waste governance in such a way as to illuminate experiences in different cultural contexts; what is termed here a geographically sensitive approach.

The geographically sensitive approach is presented in Part 2 and comprises a tripartite framework that considers policy interventions, interactions between actors in relation to those policy interventions and finally the outcomes of those interactions and interventions embedded within an appreciation of social and economic context. Two countries, Ireland and New Zealand, are used as case studies to test this governance framework because, in spite of the obvious geographical distance, there are a number of similarities in historical context and social structure between the two nations. Chapter 4 introduces the comparative analysis by providing an overview of each country's political, economic, social and environmental context as the foundation for an examination of the waste governance landscape. Attention is also paid to the methodological issues relating to the comparative governance analysis. Chapters 5 (Ireland) and 6 (New Zealand) detail the shape of the waste legislation through a consideration of waste policies in each country including attention to the actors and agencies involved in the delivery or reception of those policies and a discussion of the mechanisms or tools development to deliver them. This is followed by an examination of the interactions and relationships between the different tiers and spheres of waste governance that help to shape and reshape the waste governance landscape previously identified. A space is then created for consideration of what Hajer (1995) has described as the argumentative struggle that occurs when actors attempt to frame debates according to their interests whilst also characterising the discourses of other actors in particular ways. By focusing on outcomes the final section of Chapters 5 and 6 examine the culmination of policy landscapes and interactions between tiers and spheres of governance. Outcomes, in this context, are considered not only as the resulting impact of policies and programmes in terms

term and with respect to more fundamental definitional issues. Within the field of political science for example Kjaer (2004, 3) illustrates the diversity of definitions by contrasting the work of Rhodes (1997), Rosenau (1995) and Hyden (1999)

> Governance refers to self-organizing, interorganizational networks characterized by interdependence, resource-exchange, rules of the game, and significant autonomy from the state (Rhodes 1997, 15).

> Global governance is conceived to include systems of rule at all levels of human activity – from the family to the international organization – in which the pursuit of goals through the exercise of control has transnational repercussions (Rosenau 1995, 13).

> Governance is the stewardship of formal and informal political rules of the game. Governance refers to those measures that involve setting the rules for the exercise of power and settling conflicts over such rules (Hyden 1999, 185).

An explanation for the diversity in definitions that these quotations illustrate is that the definitions come from divergent sub-fields of social and political science. Whereas Rhodes is talking about governance in the sense of a process of reform occurring in public administration Rosenau is referring to the emergence of, and attempts to resolve, global problems and Hyden adopts governance as a means to engage with theories of third world development. These three areas of social science – public administration, international relations and comparative politics – have been at the leading edge of governance studies, but there are increasingly high profile interventions from geography, sociology and economics that are enriching debates by paying attention to issues of scale, society and efficiency in governing matters.

Disciplinary and definitional divergence has led a number of governance theorists to acknowledge that we 'are still in a period of creative disorder concerning governance' (Kooiman 2003, 5). Nevertheless while analyses of governance are diverse they tend to fall into two main camps. The first is focused primarily on empirical descriptions of processes that relate to the governing within contemporary society while the second adopts a conceptual consideration of the role of the state within social systems (Pierre 2000). Within the conceptual literature Peters (2000) suggests a delineation between research that examines the role of the state in influencing social and economic developments (state-centric or 'old' governance) and that which adopts a broader analysis of formal and informal associations particularly through networks and partnerships (society-centred or 'new governance').

At the forefront of analyses, empirical or conceptual, has been the question of whether there has been a shift from government to governance and if so whether the authority and power of the nation state has been eroded by such a shift. There are those who argue that there has been a hollowing out of the nation state as functions are dispersed to supranational entities, localities and to non-state actors (Jessop 1994; Macleod and Goodwin 1999). However others question whether such processes are indicative of states losing control, suggesting instead that they may reflect attempts by states to reorganize in the face of changing conditions. Fundamentally this position is based on the view that it is erroneous to conflate state structures with state power (Pierre 2000; Pierre and Peters 2000; Swyngedouw 2000). The work detailed in

Pierre (2000) in particular clearly supports the view that current trends demonstrate a process of state transformation rather than a decline in state authority. Equally it is oversimplistic to suggest that there has been a smooth, linear trajectory from a position of pure government to one of multifaceted governance. Many studies have demonstrated that government-governance practices can ebb and flow over time and across space (Jessop 2004; MacLeod and Goodwin 1999) as a result of jockeying between actors, institutions and organizations (Rhodes 2000). Nevertheless much of the work on governance concurs that the authoritative allocation of values is not the sole preserve of formal nation-state governments, rather it is dependent on interactions with and relationships between manifold institutions and actors from public and private sectors and civil society at a range of scales and at particular moments (Kjaer 2004) thus creating new 'geographies of governance' (MacLeod and Goodwin 1999, 505). Central to these geographies are governing structures operating at and across a range of scales and through networks of associations (Hooghe and Marks 2003).

In the light of these apparently new geographies of governance there has been considerable attention to the methods or mechanisms that might facilitate actors to generate mutually satisfactory and binding decisions through negotiation and deliberation (Martello and Jasanoff 2004; Schmitter 2001). These normative debates have focused on *good governance*, defined as 'rules, processes, and behaviour that affect the way in which powers are exercised ... particularly as regards openness, participation, accountability, effectiveness and coherence' (CEC 2001, 8), thereby resurrecting old debates about legitimacy in decision making and demanding discussions of power and interests in governing practices (Kjaer 2004). From a critical perspective Stoker (2000) engages explicitly with these issues of legitimacy, power and interests through his analysis of governance failure. He identifies a lower tier of governance failure, which he sees as a lack of engagement resulting from weakly defined opportunities for dialogue and negotiation between partners leading to a failure to achieve some social purpose. He also identifies a higher tier of governance failure referred to as an inability to produce 'more effective long-term outcomes than could have been produced using markets or imperative co-ordination by the state' (Stoker 2000, 105). Consideration of governance failure is useful because it draws attention to not only the practice of governance (how governance happens) but also its impacts and how those impacts are contested.

Despite the plethora of governance studies there are critical positions that question whether such a broad term can be useful in developing a coherent analytical framework (Marinetto 2003). There are also concerns that theory rather than grounded, empirical investigation of governing practices dominates the governance field. Indeed numerous authors call for more detailed empirical testing of theoretical models of governance (Eberlein and Kerwer 2004; Kjaer 2004; Kooiman 2003; van Kersbergen and van Waarden 2004) in order to facilitate a deeper understanding of the approaches and abilities of nation states to govern in an increasingly complex world. Such testing could, as Jordan et al. (2005, 477) propose, provide useful commentary on the divergent claims made about the extent and/or timing of governance transformations. The challenge for governance analysts therefore is to 'preserve the conceptual breadth of the term [governance], whilst simultaneously gaining analytical precision needed to empirically assess any relevant temporal, spatial and sectoral patterns' (Jordan et al. 2005, 478). One area where researchers are attempting to respond to this call for the co-development of theory and empirical understanding is within the field of environmental governance.

Environmental Governance

Increasingly studies of environmental issues are adopting and adapting notions of governance to help explain the complex processes of policy making and implementation that include not only the participation of national governments but also international and local government and non-government actors. The aims of these environmental governance studies have been to two-fold, first to re-scale issues that had previously been seen as either the preserve of nation state negotiation, such as global climate change, or of only local concern, such as air quality. Second to recognize the role of non-state participants in policy making, whether they be civil society organizations (Keck and Sikkink 1998; McCarthy 2005) or the private sector (Grossman 2000; Kolk 2000; Roome 1998). As Bulkeley and Betsill (2003, 9) argue, following Auer (2000), while the distinction is often made between 'global' processes and actors, and those which are 'local' such a binary opposition does not take account of the scalar transcendence of environmental issues nor the multi-dimensional nature of their politics.

Of course attention to the governing of environmental matters is not a new phenomenon. As Davidson and Frickel (2004) point out studies relating to the environment that engage with governance, broadly defined, have been in existence for the past fifty years in the social sciences. Early environmental governance research was conducted for the most part by political scientists who critically analysed the formation and implementation of environmental policy (as detailed in Sabatier 1979) while more recently conceptual developments such as risk society (Beck 1995), ecological modernization (Hajer 1995) and global environmentalism (Young 1999) have become prominent features of debates within sociology, economics and geography elevating environmental governance to what has been described as metatheory (Davidson and Frickel 2004). The metatheoretical concern of these more recent developments still focus primarily on nation states and their role in creating or resolving environmental problems although a number of authors have sought to bring other levels of government (Bulkeley and Betsill 2003) and other spheres of governance (Keck and Sikkink 1998) into the frame of analysis.

It is through examination of the scalar evolution of governance that the environmental field has been particularly innovative. For example, Martello and Jasanoff (2004) note the juxtaposition of global and local, and of the universal and the particular, in international regimes and policy discourses of environment and development, not least within the realms of sustainability. Environmental issues have sought recognition on the global stage since the 1970s with limits to growth discourses highlighting concerns about the finite carrying capacity of the earth and emphasising the interconnectedness of ecological systems. These themes were developed with explicit political articulation following the publication of *Our Common Future*

(WCED 1987) and subsequently the documents that emerged from the negotiations at the global Rio Earth Summit in 1992. Global institutions such as the World Bank and the United Nations have incorporated environmental management discourses in their operations and multilateral environmental agreements have proliferated with the support of scientific and expert bodies such as the Intergovernmental Panel on Climate Change. At the same time non-governmental environmental organizations have also globalized their campaigns, building their knowledge base and producing their own strategies for environmental protection and sustainable development that have not always coincided with the moves proposed by other global institutions. However critiques of global environmental governance analyses suggest that dominant theories emerging from international relations, be they focused on regimes or notions of global civil society, tend to underplay the role of the sub-national state. In response Bulkeley and Betsill (2003) use the development of scientific and political responses to climate change research as an exemplar of the rediscovery of the local in global environmental governance. They highlight how initial studies of climate change focused on providing evidence of impacts on social and ecological systems at a transnational scale, while newer studies have downscaled their analyses to national and sub-national arenas. The downscaling has led to the recognition of diversity in local knowledges that in turn has generated calls for greater stakeholder and community participation in environmental governance activities. The creation of locally resonant analyses has also created new communities and coalitions of interest such as the Alliance of Small Island States (AOSIS) a group of nations united by a precarious position in the light of climate change and rising seas. As Martello and Jasanoff (2004, 5) conclude, it appears that 'global solutions to environmental governance cannot realistically be contemplated without at the same time finding new opportunities for local self-expression'. The complexity and uncertainty of environmental issues also means that their governance cannot be simply assigned to technical or scientific resolutions that ignore normative issues and value judgements relating to equitable burdens of responsibility (Agarwal and Narain 1991). Yet incorporating local knowledges in decision making is not easily achieved within the confines of modern decision making systems that are replicated in global environmental regimes. Although there have been moves to introduce more participatory practices it is asserted that scientific knowledge is still seen as separate from (and frequently universal and free from subjective bias) and often superior to local understandings within global environmental governance arenas (Martello and Jasanoff 2004). It is also argued that where a local governance perspective, which by its very nature explicitly considers the role of the sub-national state, has been developed there are assumptions made that governance of environmental issues considered to be 'global' still emerges on a international stage to be imposed subsequently at the national and local scale – sometimes called the trickledown model of governance (Bulkeley et al. 2003). Whichever kind of knowledge is invoked in environmental governance it remains that those with power and resources tend to be able to define whether certain issues merit the world's attention (Martello and Jasanoff 2004).

There is a synergy here between the work on governance and environmental governance that suggest new geographies of governing are emerging across scales and in places, with coalitions developing between divergent actors seeking to bring about collective solutions to problems (Hajer 1995; Keck and Sikkink 1998; Liftin 1994). These new geographies are time and place specific in that the same scientific knowledge has been used to develop divergent regulatory decisions in countries that are economically and politically similar (Brickman et al. 1985) and it is increasingly recognized that 'social interests and relationships are every bit as critical in the formation of scientific consensus as in any other domain of human activity' (Martello and Jasanoff 2004, 337). Therefore environmental issues traditionally conceptualized as being place-specific have been embraced on the global stage while at the same time attention on the global stage has attempted to reclaim space for local conditions.

The nature of locality is commonly used and understood in everyday language to denote a geographical unit that is encompassed within a larger geographical, political or administrative entity such as the nation state.[1] In the arena of environmental governance however Martello and Jasanoff (2004) argue that the category of local does not necessarily have to be grounded in particular places, but can be equally assigned to communities of interest that are bonded through institutions or united through common histories or knowledges such that it might more accurately be described as glocal (Fagan 2004). Central to this glocal reconceptualization of environmental governance is a challenge to the traditional view that policies simply trickledown from one level of government to the next at which point non-state actors have the opportunity to influence them. The challenge to this trickledown model proposes that relationships can and do emerge between diverse actors operating across non-contiguous scales through networks. Networks of local authorities, for example, may seek to influence debates about climate change at the European level without necessarily going through nation state channels to do so (Bulkeley et al 2003).

The identification of networks, networking and partnerships as 'sites where governance can, and does, take place' (Bulkeley and Betsill 2003, 18) is another key area of convergence between governance and environmental governance literature and there is a general acceptance that networks are a 'defining characteristic of the new [environmental] governance' (Leach and Percy-Smith 2001, 30). Here the contribution of environmental governance is an extension of the network approaches developed by Rhodes (1997), Castells (1997) and Jessop (1995), amongst others, particularly in the light of EU environmental policy analysis where states are seen as 'one among a variety of actors contesting decisions that are made at a variety of levels' (Hooghe and Marks 1997, 23). The argument that environmental governance is not strictly ordered or hierarchical allows for the possibility that non-state actors and institutions can be active in seeking partnerships or associations to influence, even propose, policies at a range of scales from the sub- to the supra-national. According the O'Riordan and Church (2001, 22) this view of governance 'offers opportunities and threats to various social groups, depending on their access to resources and support, and on their collective capacity to identify and accommodate change' as policy networks differ in structure and nature and not all nodes within networks have equal influence over policy outcomes. Marsh and Rhodes (1992) and

---

1   The contested nature of scale is an area of much deliberation in the academic community as documented in Whitehead (2007).

O'Riordan and Jordan (1996) argue that networks can be placed along a spectrum. At one extreme is a tightly organized policy community based on a single department alone with a loose network at the other where a broader range of actors, that engage with government departments on a less predictable basis, come together over issues where there is a significant degree of contention. Nonetheless across this spectrum is the assumption that for the achievement of policy outcomes with a minimal level of conflict governments need 'the assistance and co-operation of other groups' (Smith 1997, 35). Networks are also seen as a manifestation of the diffusion of governance away from the nation state to both the global and local levels and as a site of governance innovation in their own right with networks acquiring authority by generating and disseminating knowledge as information, ideas and values (Bulkeley and Betsill 2003; Lipschutz 1997).

Policy networks have been critically analysed by a range of authors in terms of how networks are formed and how policy change occurs through them. Most pertinently difficulties are demonstrated where, with issues such as climate change, there are many government departments implicated and a range of actors with varied interests involved. Bulkeley (2000a) suggests, for example, that rather than a single network within climate change policy there are actually layered, intersecting networks operating which demand more nuanced analysis than simply asserting the existence of networks. This would require a detailed understanding of the actors and institutions involved, the ways in which they attempt to govern and the outcomes of those governing practices. While networks have been promoted as mechanisms to create consensus on governing through negotiation they have also emerged as channels through which systems of governance can be resisted when environmental resources have not been protected (Keck and Sikkink 1998; Yearley 1995).

Initially empirical studies of environmental governance tended to focus on the global institutions of governing in particular sectors, such as climate change (Paterson 1996), ozone depletion (Liftin 1994), desertification (Porter et al. 2000) and hazardous waste (Wapner 1998). However the terrain of environmental governance studies is diversifying with an increasing number of studies examining particular actors within governance environments such as environmental non-governmental organizations (McCarthy 2005), new policy instruments (Jordan et al. 2005) and corporate environmental responsibility (Lund-Thomsen 2005) as well as the complexities of interactions between scales and spheres of governance (Jordan and Schout 2005; Klooster 2005; Thompson 2005; Vogler 2005). The environmental governance field is thus replete with empirical work and theoretical advances from a range of disciplines and sub-disciplines within the social sciences, typical of what Spaargaren et al. (2000, 2) term the 'scattered landscapes of environmental studies'. Yet there are opportunities for advancement both in terms of sectoral coverage and analytical endeavours. Most significantly are calls to expand the number of comparative studies (Jordan et al. 2005), to expand the analysis of governance to consider other aspects of environmental concern and to broaden research beyond a state-centred or society-centred approach in order to engage with an 'explicit examination of state-societal relations' (Davidson and Frickel 2004, 472). It is in response to these gaps in research that this book has been developed as comparing the governing of municipal solid waste provides an arena through which to address these three developmental avenues. However this book is not the first foray into waste governance per se as the following section demonstrates.

Waste Governance

If environmental governance is indicated by multiscalar interactions of different actors from a range of state and non-state organizations then the management of waste is no exception. Waste is already a matter for global environmental governance through the Basel Convention for the transnational movement of hazardous waste (Fagan 2004; Greenpeace 1993; Wapner 1998) and developments in recycling markets are creating new global geographies of waste flows. It is now common to find national waste management plans in western industrialized countries and the EU has created numerous supra-national waste directives and policy statements. The private sector is an increasingly familiar actor in the delivery of waste services across the globe and civil society organizations are attempting to exert an influence in waste policy formation and implementation. Equally the spheres and tiers of waste governance are becoming intertwined through public private partnerships and networks of waste organizations such as the International Solid Waste Association (ISWA) are attempting to represent the interests of participating institutions and perspective in policy debates. At the same time transnational advocacy networks such as the Global Alliance for Incineration Alternatives (GAIA) and the Zero Waste Alliance seek to support localities in attempts to move towards zero waste to landfill and in opposition to incineration technologies (Davies 2005). So while traditionally seen as a matter of disposal for the local state, the production, transport and treatment of waste, with its identified impacts in social, political, economic and environmental realms, is now also an issue of concern for nation states and supra-national entities, the private sector and civil society.

If the landscape of waste governance thus described is nuanced then academic analyses of that governance are becoming equally variegated. From a position of relative marginalization there are a number of authors developing governance frameworks for the specific analysis of waste. Some have emphasized the role of networks in the functioning of waste practices (Fagan 2004), while others have developed arguments using institutions as the driving force in waste management systems (Parto 2005). While networks and institutions are familiar features of governance analyses there are also studies that have expanded these conceptual boundaries to engage with Foucauldian inspired governmentality (Bulkeley et al. 2005) and political ecology in order to understand how waste is governed (Myers 2005). These approaches to analysing the governing of waste are considered below.

Networked Waste Analysis

The networking approach takes scalar concerns as its point of departure for analysis. At the heart of this approach is recognition of increasingly fluid interactions between cultures, societies and economies across the globe that suggest a localization of the global and the globalization of the local (Dirlik 1999) in many facets of social life including the regulation and management of waste. As a key proponent of this approach Fagan conceptualizes waste as an emblem of modernity, as a 'globally circulating fluid, its production and management governed well beyond the nation state' (2004, 87). A useful contribution of the networked approach to governance analysis is the acknowledgement that while regulatory environmental frameworks have supranational dimensions, as mentioned previously in relation to the Basel Convention and the EU, they are also grounded in the micro-processes of everyday life. In this way individuals are actively implicated in social practices that result in the creation of waste through consumption patterns which then connect consumers with locally organized networks of waste collections, be they publicly or privately operated. It is this explicit connection of waste with society, bringing social theory into the realm of waste management and situating it within the policy context of waste management that characterizes the networked waste governance approach (see also O'Brien 1999a; 1999b; Yearley 1995).

Within the networking approach all relationships between different waste actors and institutions are characterized as networks, the EU is described as a 'network made up of nation states' (Fagan 2004, 91). The nation state is reconfigured, following Carnoy and Castells (2001), as a network state typified by bargaining and interaction with other players in decision making. While local communities resisting the development of waste infrastructures are identified as 'clearly networked to the global environmental condition' (Fagan 2004, 97) through transnational advocacy associations. Building on the ideas developed by Martello and Jasanoff (2004), a networked waste governance approach moves away from a dualistic view of the local as peopled and the global as the terrain of capital through this concept of glocalization, or networked political action. As with broader discussions of environmental governance it is argued that there is a need to extend waste governance analyses beyond simply identifying existing networks to investigate the capacities of these networks, and more specifically the influence of actors within those networks. Although more work needs to be conducted in relation to waste networks Fagan (2004, 100) concludes that the 'neo-liberal discourse in relation to waste is no doubt a dominant one'. So while there may well be networks of waste in the new geographies of waste governance analysts need to look more carefully not just at participation in governing networks, but how that participation is structured through institutions.

Institutional Analysis of Waste

Parto (2005) argues for a more explicitly institutionalist analysis for waste on the basis that by analysing the structuring factors of waste governance – that is the institutions that shape the waste policy process – a better understanding can be achieved of why policy outcomes (be that volumes of waste produced or levels of recycling) often fail to meet the expectations or intentions enshrined in policy documents, statements and directives (such as sustainable waste management). Parto (2005), like Fagan (2004), begins by highlighting the limitations of nation state politics for the resolution of contemporary problems, such as waste management, drawing on Hajer's notion of power diffused through 'transnational, polycentric networks of governance' (2003, 175). He also acknowledges the role of networks for governance at the supra-national scale, but in contrast to Fagan (2004) he emphasizes the way in which institutions within those networks operate. Where Fagan (2004) sought to bring society into the technical field of waste management Parto is more explicitly focused on waste policy analysis, an activity that he asserts cannot be done meaningfully 'without addressing issues of governance and accounting for the role of institutions' (2005, 2). Here he engages with conditions of conflict that surround the process of governing such that governance is defined as 'the contestations around how resources are actually allocated' (Parto 2005, 4). It is how actors organize themselves through formal (e.g. family, corporations, trade unions and the state), semi-formal (rules, conventions and models) and informal (norms, habits and customs) institutions to structure that contestation that is the key to waste governance under this institutionalist framework. As with the networked waste governance approach, there is a long and diverse history to the theoretical evolution of institutionalism which Parto (2005) traces back from Jessop (2001) and Scott (2001) to Marx and Durkheim, but from the mass of definitions and categorizations of institutions he establishes three areas of emphasis: scales or territories of governance, levels of interaction amongst and between individuals, organizations and society, and finally systems or spheres of society (such as social, economic, political or ecological). This scales-territories-systems perspective is used to create a typology of institutions along a spectrum of formality from informal behavioural norms, through cognitive structures and associative institutions (that facilitate certain prescribed and privileged interactions between public and private interests) to regulative institutions and finally formal constitutive institutions that set the bounds of social relations. By using this typology to identify different types of institutions involved in waste policy Parto (2005) suggests a greater understanding of why governance operates in the way that it does in specific contexts and at particular moments. Methodologically Parto (2005, 13) calls for an eclectic qualitative model (Nelson and Winter 1982) that facilitates historical examination of the field, allows mapping of institutional evolution and interaction over time and engages directly with participants in policy communities. These interpretative endeavours consider contextual details and identify informal institutions. In sum Parto (2005) presents a detailed structure for the analysis of waste governance which facilitates the kind of comparative empirical analysis called for by analysts of waste governance such as Jordan et al. (2005). However despite his recognition of the plurality of governing sites and their evolution over time Parto (2005), like Fagan (2004), tends to emphasize the structures of governance rather than the processes of governing. It is concern with these processes that define more neo-Foucauldian approaches to waste governance.

Modes of Governing Waste

In their work on municipal waste management in the UK Bulkeley et al (2005) seek to address both the complex structures and processes of governance simultaneously while recognising the plurality and multiplicity of governing sites and activities through an analytical 'modes of governing' approach. Their approach 'engages with rationalities, agencies, institutional relations and technologies of governing that coalesce around particular objectives and entities to be governed' (Bulkeley et al.

2007, 1). The aim of such an approach is to illustrate the dynamic and multifaceted nature of waste governing as well as describing the modes of governing that currently shape the policy and practice of waste management.

As with Fagan (2004) and Parto (2005), Bulkeley et al. (2007) construct their analytical framework in the light of empirical findings. The methodological approaches used by all these authors have commonalities in their adoption of documentary analysis and qualitative interviews as the baseline means of accessing experiences and activities of waste governance. Bulkeley et al. (2007) are however more expansive in their empirical design with the introduction of workshops and detailed small-scale case studies of waste practices. It was through this empirical work that the authors found existing environmental governance frameworks insufficient to account for the experiences they uncovered in the waste sector, both in terms of facilitating an engagement with the processes through which governing occurs and for considering the relationships between policy and practice. In part this insufficiency is the result of a tendency to simply overlay theoretical frameworks of governance – be that networking or institutionalism – onto specific environmental sectors such as waste without a detailed consideration of whether the nature of waste and the way that it is governed is distinctive. However Bulkeley et al. (2007) also argue that the bulk of work in the governance arena has been focused on defining what governance is (or is not) rather than creating a deeper understanding of how governance takes place.

It is to neo-Foucauldian theorists that Bulkeley et al. (2007) look for more clarity in the realm of 'how' governance works, that is to understand the means through which policies and programmes are enacted (see also Dean 1999; MacKinnon 2000, 295). They do this by employing the notion of governmentality, interpreted as activities that aim 'to shape, guide or affect the conduct of some person or persons' (Gordon 1991, 2). Like Parto (2005) there is a recognition of contestation in this interpretation of governing that constantly seeks to define *what* should be governed (objects of governance) and *how* those objects should be governed (the nature of that governance), what has been called 'the rationalities of governing' (Bulkeley et al. 2007, 8). These rationalities are materialized through various programmes for action, be they policies (such as regulations, market mechanisms or indicators) or infrastructures (including pay-by-use waste collections, recycling centres or civic amenity sites), which are together termed governmental technologies. It is through these technologies that governmental rationalities are disseminated, political authority determined and subjectivities established (Raco 2003) and it is analysing these processes together as 'regimes of practice' that Bulkeley et al. (2007) identify as a key pathway to understanding how such policies emerge, persist or evolve. Essentially a governmentality approach allows the focus of attention to be placed on how regimes are governed (Dean 1999), rather than simply describing institutional arrangements. Yet, following the concerns of Foucault with the microphysics of power, adopting a governmentality lens should not obscure the ability of actors to challenge, resist or reinterpret those regimes (Herbert-Cheshire 2003; Raco 2003). Given these caveats attention to the context of regimes that appreciates scalar dimensions of governmentalities is essential (Uitermark 2005).

While different in structure and emphasis the conceptual approaches of governance and governmentality detailed above all share common points of departure in their desire to question a homogenous view of the state in contemporary society, but in essence governance debates draw attention to the complexity of institutional contexts (and that is institutions in their widest sense) while governmentality frameworks add richness to understandings of how governing practices occur. Bulkeley et al. (2007) attempt to bring together the benefits of these approaches while minimising their limitations through a 'modes of governing' approach. A mode of governing is defined as a 'set of governmental technologies deployed through particular institutional relations through which agents seek to act on the world/other people in order to attain a distinctive objective in line with particular kinds of governmental rationality' (Bulkeley et al. 2007, 13). Empirical work conducted by Bulkeley et al. (2007) identified the co-existence of multiple modes of governing waste from disposal through diversion to eco-efficiency and waste as a resource. Each of these modes has associated components that are articulated through different governmental rationalities (that is aims and objectives that are manifest through policies and infrastructures) and governing agencies (such as local authorities, the EU, waste contractors or non-governmental organizations). Governmental rationalities and governing agencies are then associated with particular forms of institutional relations (hierarchy, networks or community), governmental technologies (weekly bin collections, recycling or reuse practices) and governed entities (local authorities and individuals as passive or active citizens). Conceptualizing modes in this way facilitates analytical clarity although it is recognized that modes of governing are not discrete entities but are themselves fluid structures and processes as the technologies associated with them evolve and institutional relations change. Seen through this analytical lens waste has become problematized at a range of scales from the European to the local and different rationalities and technologies for governing have evolved in particular locations which in turn leads to a variety in the ways in which waste is [re]conceived by individuals and public, private and civil society sectors. Hence context is key to understanding both how and why governance takes the forms that it does. The modes of governing approach then moves some way to answering Davidson and Frickel's (2004) call for more attention to state-society relations, but its application needs to be extended to different contexts to establish the degree to which differences emerge across space. The studies so far described all refer to the governing of waste in western industrialized countries, specifically European contexts, but concerns with waste are not confined to such locations. In less economically developed contexts a political ecology approach has been proposed to examine waste governance.

Political Ecology of Waste

The work of Myers (2005) and his articulation of waste management in urban areas of Sub-Saharan Africa adds geographical and conceptual diversity to waste governance analyses. He adopts a political ecology framework for examining the impacts of solid waste management under the auspices of sustainable development planning. He does this specifically through an analysis of the development of the United Nations

Sustainable Cities Program that aims to assist cities in the collection and disposal of waste. As with the other analytical approaches detailed in this chapter political ecology studies have been subjected to criticism. Despite an increasing diversity of political ecology analyses (see for example Zimmerer and Bassett 2003, 3) there have been charges of insensitivity to the complexity of scale and an overemphasis on local and rural contexts (Schroeder 1999). However Myers (2005) moves some way to addressing these concerns by considering the interactions between the Sustainable Cities Programme, a global initiative organized by the United Nations, and selected cities in Urban Africa.

Rather than focusing on networks, institutions or modes of governance per se Myers (2005) structures his arguments around four themes that span economic, environmental, political and cultural spheres under the titles of neo-liberalism, sustainable development, good governance and the politics of cultural difference. However his overarching aim is the examination of outcomes. In this way his work can be seen to complete the triad concerns of governance that is the how (the structures and practices), why (rationalities) and what (impacts or outcomes) of governing processes. In particular Myers (2005) brings more detailed attention to history, culture and society in a field dominated by political science, economics and technical assessments of environmental degradation caused by the globally familiar process of wasting. Initially developed in response to an unease with the predominantly physical analyses of environmental degradation political ecology in the waste context serves to bring the ecology back into debates, 'in effect to merge cultural geography or cultural ecology with the political and economy in order to study the environment' (Myers 2005, 13). The major contribution of political ecologists has been to highlight the politicized nature of environmental issues and the power inequalities that shape decisions about environmental management or exploitation (Bryant and Bailey 1997; Hardoy et al. 1992). Ultimately Myers (2005), following Pelling (2003) and Gandy (2002), concludes that waste can be perceived as a resource, or hazard, or simply ignored depending on the character of government-society relations and it is the interface of those perceptions with bureaucratic structures and cultures that influences the decisions and outcomes of waste governance.

While the preceding discussion demarcates distinct approaches to waste governance analyses in as much as the work can be associated with the broader analytical categories of networking, institutionalism, governmentality and political ecology respectively there are also areas of commonality. For example all approaches adopt a predominantly qualitative methodology for capturing the practices of governing and all are concerned to identify the manifold participants in governing practices. It could even be argued that the approaches differ more in terms of emphasis than in substance. Both the institutional and networking analyses tend to focus on highlighting the limitations of the nation state in governing waste while governmentality approaches are concerned as much with processes as with structures of governance. Finally political ecology of waste seems to prioritise a reunification of concerns with ecology and culture within the political realm of decision making in order to evaluate the outcomes of governance.

Conclusion

This chapter has presented key dimensions of debates within the field of governance and its sub-fields of environmental and waste governance. It has been established that there exists a general consensus across governance analyses (generally as well as in the environmental and waste arenas) that attention needs to be paid to the institutions, structures and actors involved in the process of governing. There is less consensus however about whether to focus on the role of the state in influencing social and economic development (state-centred analysis) or to take a broader approach that draws attention to both formal and informal associations, networks and partnerships (society-centred approaches). However both state and society centred approaches admit that the authoritative allocation of values is no longer the sole preserve of formal nation state governments and that this has created new geographies of governance such that 'the role of the state is not governed by some determinate and finite notion of capacity, but rather through negotiations in which actors and institutions mutually define their respective roles' (Bulkeley and Betsill 2003, 191).

There seems then to be space of convergence here within which it is possible to develop a state-society governance approach that both acknowledges the state as an important site of activity with regards to governance matters, but that can also accommodate attention to the influence of non-nation state actors in governing processes across scales, in places and through new associations such as partnerships and networks. Sensitive attention to these institutions, structures and actors and their complex interactions allows the construction of a detailed map of what happens (and who participates and to what extent) in governance matters, but it does not provide much of an insight into the mechanisms of governance, that is the practice and processes (or rationalities and technologies) of governing. Attention to these 'regimes of practice', that is the rationalities and technologies of governing, allows for greater comprehension of how the identified institutions, structures and agencies of governance become articulated and interact. The final element required in governance studies is a consideration of outcomes. While it is important to establish the 'what' and 'how' of governing in particular places it is also necessary to consider the outcomes of those governing moments and how governed entities react or resist the modes of governing that are being implemented. This is no simple task, particularly if analysis is to incorporate a comparison across nations and the following chapter illustrates just how wide is the range of different contexts in which waste governance practices have been considered.

It is proposed that an analytical framework which seeks to both preserve the conceptual breadth of the term governance and the insights of governmentality while maintaining some analytical precision in terms of empirical assessment will facilitate a rich understanding not only of what waste governance is, but also how the processes of governance operate and the outcomes that are subsequently produced.

Chapter 3

Garbage Governance in International Context

Introduction

The previous chapter outlined key conceptual and theoretical approaches that have emerged in relation to the examination of environmental and waste governance. It detailed the tensions between studies that have adopted an environmental governance perspective as a descriptive tool to outline how practices of governance operate and those that have focused on the development of abstract models of governance with the ultimate goal of establishing normative evaluations of different governance systems. The chapter also demonstrated that while governance studies in the waste field are increasing, they remain overshadowed by attention given to other environmental sectors such as climate change, ozone depletion, and nature conservation.

However it would be erroneous to suggest that waste has been entirely neglected as arena of study. It is the function of this chapter to illustrate that, while explicit governance perspectives on waste are rare, there are studies that have examined elements of waste management in specific localities, regions and nation states throughout the world. These studies illustrate the diversity of contexts and challenges that face the governors of waste into the 21st century and also the range of solutions that have been applied to resolve those challenges. Interestingly, despite the diversity in experiences there remains a surprising commonality in terms of overarching discourses in the waste management field. From Tanzania to the Netherlands and from Massachusetts to Delhi references are made to integrated solid waste management and the waste management hierarchy. This chapter traces the peculiar combination of diversity and commonality in waste management research through the examination of illustrative studies conducted in a range of locations across Europe, USA, Asia and Africa. These case studies are not intended to be representative of waste management practices across the globe. Such a task is beyond the scope of this book, nor would it necessarily reveal the intricacies of governing on which this book is focused. The cases do however exhibit key characteristics of governance systems – such as multiscalar interactions, networking, partnerships and conflict – identified from theoretical studies in Chapter 2. Before examining the geographically diverse studies of waste management it is useful to illustrate more clearly the link between the conceptual concepts of networking, partnerships and conflict and their practice in empirical contexts.

problems caused by increasing volumes of waste and large numbers of unregulated landfill sites with attendant environmental, health and safety concerns stimulated a government decision to replace all unregulated dumps with a rationalized system of large-scale regional landfills, financial assistance to local authorities to facilitate movement of waste to the new landfill locations and the introduction of energy recovery from incineration. The regionalization of landfills was intended to reduce environmental pollution and provide facilities for the majority of the population in standardized operations but the plans failed to set detailed timescales for the closure of unauthorized landfills or schedules for the construction of the new facilities. The statutory process for planning of new facilities, requiring approval at local, district and national levels, also caused delays with significant local opposition to the landfill developments. Nissim et al. (2005, 324) reported that 'as a result, detailed plans for authorized sanitary landfills were not approved in most areas and over two thirds of Israel's population remained without a comprehensive solution to the problem of solid waste disposal'. Resistance was such that the Ministry for the Environment was forced to develop new means of conflict resolution appropriate to the specific concerns of the local communities. In one location, Talya, in the northeastern part of Israel a collaborative conflict management approach was developed where adjacent communities were invited to participate in the complete environmental impact statement review and supervision of waste transport and landfill operations. In Dundaim, located near Beer-Sheva in the centre of Israel, opponents of the landfill were offered a 'host fee' whereby the regional council would gain a certain amount of money per tonne of waste disposed at the site. This solution did not placate all opponents however and residents from nearby settlements used a variety of legal means to try and prevent the facility operating. Although not all conflicts surrounding waste management in Israel were resolved by 2004 the Ministries of the Environment and of the Interior had succeeded in closing or improving 50 per cent of the unauthorized dumps and rehabilitated ten sites following closure according to heightened environmental standards and are working to implement an integrated solid waste management strategy (Nissim et al. 2005).

The above examples of networking, partnerships and conflict that incorporate agencies and actors working at a range of scales from the grassroots to the supranational illustrate how features of theoretical debates about governance can be identified even in research undertaken to primarily to describe waste management practices. The remainder of the chapter examines more closely empirical attempts to examine waste governance in different locations around the globe with case studies drawn from Europe, the USA, Asia and Africa. The cases refer to contrasting social, political, economic and environmental contexts, which is reflected in the nature of the problems and solutions provided by various waste governance systems.

Europe: Unity in Diversity

The management of waste within European countries, even amongst member states of the EU, displays all the variety that might be expected from a culturally and economically diverse region. There exists a large body of technical research and policy analysis of waste management mechanisms and perhaps more than in any other region examined here the analysis of waste governance within Europe is the most extensive. It should be noted however that this has not led to the emergence of an agreed model for waste management, beyond broad acceptance of EU Directives by member states, but rather a proliferation of often contrasting approaches to meet agreed ends. In one study of European waste structures conducted by the Green Alliance (2002, 1.7) there were 'differing interpretations of what constituted municipal and household waste and different ways of measuring performance', which led to problems of comparability. Nevertheless there are two broad camps within waste governance analysis across Europe that can be distinguished, those that see the persistence of difference amongst nation states shaped by unique intersections between institutions of governance and those that identify a 'Europeanization' of waste policy and a convergence of approaches driven by EU Directives.

Despite the emergence of a highly developed legislative framework for the environment within the EU the differences between cultures, modes of governance, economic development and administrative structures of member states persists and generates what Parto (2005, 2) has termed 'conflicting perspectives and competing agendas'. Indeed the EU's notion of *good governance* (CEC 2001) has sought to support networks of governing that include, but also move beyond formal government structures allowing for 'self-organization' within its dictats (Schout and Jordan 2003).

Parto (2005) illustrates the existence of diversity in governance through his institutional policy analysis of waste management sub-systems in the UK and the Netherlands. He begins by contrasting the different modes of government in the two states with the UK characterized as a strong central state with a suppressed local level of government where partnerships (mostly public-private partnerships) are seen as the result of the sub-contracting of state functions, what Jessop (1999) would call the 'hollowing out' of the state. However civil society organizations and associations continue to operate under competitive conditions for scare financial resources from the state and non-state bodies (Fairbrass 2003). As a result Parto (2005, 14) concludes that conditions of governance in the UK are shaped by 'limited social-democratic resources and entrenched neoliberalism'. In contrast he suggests that strong local government characterizes waste governance in the Netherlands, with local government working with a certain degree of autonomy for implementing policy and central government taking the lead role in defining strategic policy objectives. While the non-state sector participates in service provision and civil society organizations enjoy guaranteed levels of government funding their activities are highly regulated. The Netherlands system of governance is then defined by 'abundant social-democratic resources and regulated neoliberalism' (Parto 2005, 15).

While both countries aspire to integrated waste management practices that emphasize prevention and waste reduction as preferred options, followed by reuse and recycling then incineration and landfill disposal, there are different interpretations of how best to attain this common goal. The role of incineration is a particularly stark example of this difference. Within the Netherlands incineration provides a significant means of waste disposal and it has become an institutionalized and publicly accepted waste management tool. In contrast waste actors and the public in the UK are less united in their appreciation of incineration as a suitable mechanism to resolve waste management challenges. Parto (2005) suggests that the differential response can be linked to the ways in which the state reacted to concerns over the health and safety impacts through dioxin emissions of early incineration processes during the 1960s. He contrasts the 'deny and defend' approach of the UK government with the 'coalition model' adopted in the Netherlands where the government was able to diffuse confrontation through the inclusion of multiple stakeholders and diverse positions with regards dioxin emissions in the process of problem resolution. The remaining conflicts over incineration in the UK have led to the persistence of landfilling as the primary route for waste with almost 80 per cent of municipal waste being managed this way compared to only 20-35 per cent in the Netherlands (DEFRA 2004). This difference cannot be reduced to the conflict management approach of nation states alone however and it is influenced by specific geographical, cultural and political conditions. The UK has, for example, traditionally had access to sites for landfill development as a result of mining and other activities which, allied to relatively low landfill taxation rates compared to the Netherlands, means there is little economic incentive to look to alternative mechanisms. Equally the level of public awareness and demand for recycling and waste minimization is low in the UK compared to other European countries and despite resistance to the siting of landfills in many locations they are often perceived to be a lesser evil than incineration. Finally, although institutional mechanisms for waste management are changing (Davoudi and Evans 2005) the responsibility for waste management is fragmented leading to a complex relationship between collection and disposal authorities (Bulkeley et al. 2005). In contrast to the UK (at least in the industrial sector) the Netherlands has managed to decouple waste production from economic growth and has devised strict criteria for the prevention of pollution through incineration, waste minimization and by supporting increased producer responsibility legislation.

Parto (2005) suggests that the differences between the two waste systems can be explained fundamentally by the contrasting institutional approaches to governance. He claims the system of majority voting in the UK detaches the process of governing from the populace whereas proportional representation as practiced in the Netherlands works to reflect the political positions of the public more closely. He uses the work of Beaumont (2003) on poverty elimination strategies to illustrate how the first past the post system in the UK creates a strong central controlling state that reverts to partnerships only as a means to try and build trust when communication breaks down between governments and publics. The Dutch model, which is also characterized by central control, in contrast seeks to actively engage publics on a more consistent and collaborative basis to resolve contentious issues. Conditions of trust between publics and waste operators – characterized by interactive policy-making – are seen as pivotal to smooth operation of waste management practices particularly as most incineration plants are controlled by central government. Responsibility for goal setting and implementation of waste management strategies is spread across national, regional and local governments and interactions between scales of government are seen as part of a collaborative effort to produce more sustainable waste management practices (Green Alliance 2002). The politicization of environmental concerns amongst citizens in the Netherlands during the 1960s and the proportional representation system of voting allowed the environment to become a feature of elections and stimulated legislative arrangements to address the environmental concerns. As a result certain environmental practices have become institutionalized to the extent that 'recycling is less "what ought to be done" and more what is done' (Parto 2005, 19). This position is reinforced in Dutch society by cultural referents that characterize waste and 'wasting' as morally wrong, but it has also been assisted by the close participation of civil society groupings in policy partnerships facilitated by central government funding. In the UK non-governmental organizations are often marginalized in favour of public-private partnerships and frequently have to compete with each other for scarce funding (Luckin and Sharp 2004).

The Netherlands have indeed achieved high levels of recycling through a combination of high landfill taxes, landfill bans and limitations on incineration capacity and a range of voluntary agreements alongside measures to support the development of markets for recyclates (Green Alliance 2002), yet challenges still remain. Levels of recycling have reached a plateau and new mechanisms will be required to continue reducing the amounts of waste being generated. High landfill taxes and landfill bans have been influential, but they were slow to create an impact initially as recycling facilities (predominantly privately operated) did not develop as quickly as had been anticipated. Equally while interactive policy making led to good communication between actors critics of the system have identified problems of complexity and compromise in legislation. Some NGOs in the Netherlands are also concerned that greater volumes of waste might be exported under EU legislation that permits the transfer of waste for recycling purposes. The concern is that as the waste market becomes increasingly internationalized there could be a rise in 'eco-dumping' that would be hard to control (Green Alliance 2002).

The diversity between the UK and the Netherlands is not unique within a European context as the research of Wilson et al (2001) and Green Alliance (2002) has demonstrated. Yet there are authors who claim that diversity of governance within Europe is declining with a movement towards Europeanization of policy (Davoudi and Evans 2005). Certainly it has been established that within Europe discourses of Integrated Solid Waste Management (ISWM), with ISWM being the attempt to systematically manage waste in an environmentally, socially and economically sustainable way (White 1996), are commonplace. As delineated in the introduction ISWM frequently involves the articulation of a waste hierarchy as a guiding framework (Green Alliance 2002), however under this common veneer lies a complexity of approaches to waste and more careful analysis of the various modes of waste governance needs to be undertaken.

USA: Delegation and Volunteerism in a Federal System

The issue of waste in the United States of America has received attention from a diverse range of disciplinary backgrounds. Alongside the more familiar technical appraisals of waste management, as discussed by Tammemagi (1999), there are a host of texts concerned with the waste by-products of a highly consumerist society (de Graf et al. 2002; Royte 2005; Strasser 1999), the justice implications of managing waste (Bullard 2000; Camacho 1999) and the historical evolution of waste management strategies (Louis 2004; Melosi 2001; 2004). In addition there is an evolving sub-discipline of waste archaeology that has engaged with the excavation of waste (Rathje and Cullen 2001) and even novels that have taken the theme of waste as a central theme (DeLillo 1997).

While there is attention to the politics of waste in many of these diverse texts, and specifically in the work of Rasmussen (2000) and McAvoy (1999), there is little explicit articulation of a governance approach to waste management that is found in Europe. This is despite a range of analyses that suggest it is governance – through privatization (Eggerth 2005) and the action of civil society actors (Walsh et al. 1997) – rather than simply formal government that plays a pivotal role in the management of waste in the USA. Even when governance has been the focus of attention, in most studies the emphasis is on the state rather than national level (Hill et al. 2002). This concentration on the state level is unsurprising as while the Environmental Protection Agency (EPA) Office of Solid Waste (OSW) regulates waste management under the Resource Conservation and Recovery Act (RCRA) it is individual states that adopt Federal standards and operate their own waste management programmes.

Authors such as Louis (2004) and Melosi (2004) emphasize the importance of the historical evolution of waste management policies for any analysis of current practices in the USA. In particular they cite the allocation of solid waste management to local level government in the nineteenth century and the development of engineering based solutions emerging from city sanitation departments that relied on municipal dumps as pivotal in shaping waste management strategies nationwide. However it is the Resource Conservation and Recovery Act of 1976 that still defines waste management practice in the USA today. In response to rapidly rising volumes of waste – estimates suggest that volumes increased more than five times the level of population growth between 1920-1970 (White et al. 1995) – the Act developed new standards for landfill sites that forced the closure of open dumps and demanded a regional approach to waste management planning. The closure of many municipal sites created what Louis (2004) calls a 'garbage crisis' during the 1990s that led to the increasing participation of private companies in the management of waste. He reports that in some places, such as Allegheny County, Pennsylvania, 100 per cent of landfills and materials recovery facilities are privately operated. Trans-state movements of waste as a result of this privatization caused problems between administrative units. Those states and counties that found themselves to be net importers of waste wanted to be able to control the amount of waste entering their jurisdiction and also to dictate the destination of that waste. Such 'flow control', as it was called by the private waste industry, led to federal litigation and eventually US Supreme Court intervention and a ruling that under the Commerce Clause of the US Constitution municipal solid waste was protected from such restrictive practices by states. This decision fuelled the privatization of waste even further. According to Federal and State laws municipalities still predominantly manage waste, but the daily operations dealing with waste are increasingly dominated by a small number of large private companies responsible for waste collection, recycling and disposal.

To summarize, Federal Government is responsible for environmental legislation and through the EPA a set of national minimum standards are established for all states. States however retain a significant amount of independence and can, in addition, develop laws for the management of waste as long as the national minimum standards are respected. States are also responsible for permitting, monitoring and enforcing waste management activities. This allows for flexibility within states as to how standards are met and what targets are set above those minimum standards. It also, as a result, means that there is considerable variation between states in terms of practice. In California, for example, the management of waste is conducted through the California Integrated Waste Management Board (CIWMB) that has developed and implemented a number of programmes for waste management through the Integrated Waste Management Act established in 1989. One of these was to adopt Zero Waste partnerships with local government, industry and the public. Programmes are carried out by Local Enforcement Agencies with 531 jurisdictions having the power to tax, enact and enforce local legislative requirements (Hill et al. 2002). California has achieved a good level of recycling with its waste management strategy focused around a mandatory 50 per cent target for diversion of solid waste with strict penalties for non-compliance. While the ownership of disposal facilities is shared between public and private agencies the majority of collection services are privately operated. This mix, in conjunction with the flexibility of the Integrated Solid Waste Management Act regarding how the 50 per cent target is met, has generated innovative solutions and contracts for waste management in the state. There are still disagreements however about how to measure whether these recycling targets have been met or not.

As in California the management of waste in Massachusetts is conducted through a central state body, the State Department of Environmental Protection that works along with local Boards of Health to regulate collection and disposal of waste. The Department also produces the solid waste plan for the state in accordance with Federal law. Under the state tier of government there are over 350 municipalities that have the power to enact and enforce local legislation and to tax waste activities to a limited extent (Hill et al. 2002). In contrast to California however Massachusetts have set voluntary goals preferring to adopt an incentive-based system of waste management rather than a reliance on regulatory mechanisms although a number of landfill bans on specific materials have been introduced. The bans worked in conjunction with moratoria on landfill and incineration capacity over the past decade to make recycling a more attractive option within the state. A Bottle Bill, that directs money from unredeemed bottle deposits to recycling and hazardous waste clean-up programmes, supports recycling initiatives within local authorities and pay-as-you-throw schemes, has been introduction in most municipalities. Recycling rates have improved as a result although there is concern that the focus on disposal bans deflects attention from waste generation and product stewardship initiatives based on voluntary agreements are being investigated.

The delegation of responsibility for implementing waste policy is not uncommon for industrialized countries, but it can lead to tensions between scalar priorities as identified by Rasmussen (2000) in the USA. Local regulatory agents have to balance local pressures and needs with the demands of regional and national regulators.

Scalar tensions are not the only arenas of conflict within waste management in the USA. As the wealth of literature on environmental injustice in the waste sphere indicates (see Camacho 1999) there is a widely reported view that 'garbage follows a strict class topography. It concentrates on the margins, and it tumbles downhill to settle in places of least resistance, among the poor and disenfranchized' (Royte 2005, 40). The existence of a multitude of minority groups and a stark rich-poor divide has lead to conflict particularly between communities and waste service providers over the location of facilities (Clarke et al. 1999). One particular area of complexity is the management of solid waste within native tribal communities. Alaskan Native and American Indian Tribes number more than 550 and represent sovereign nations within the USA that have a unique legal position and specific relationship to federal government (Ortiz 2003). The native tribes form a distinct level of governance in the USA. They are discrete governing entities with control over their people and land and with the capability for self-government. However Ortiz (2003) identifies significant waste management problems within Indian lands in terms of landfill sites not meeting Federal standards and particularly the co-mingling of hazardous and non-hazardous wastes. The problems that result have been examined by a number of commentators as cases of environmental injustice with the marginalization of tribal voices from policy formulation, regulatory and enforcement fora (Aufrecht 1999; Tirado 2001). Following the identification of distributional imbalances between the treatment of tribal and non-tribal communities with respect to waste management there have been attempts to develop better relationships between Federal and Tribal governments under a partnership framework. These partnership arrangements are however complicated by disagreements between States and tribal communities about how best to govern Indian country and its people.

It has already been established that while the RCRA is the primary Federal statute for waste management, most of the authority for regulating and enforcing waste treatment has been devolved to individual states, but there is a difference here with respect to how tribal communities are treated. The RCRA defines tribes as municipalities but they cannot be the delegated authority under the RCRA which has led to tension and claims of an infringement of sovereignty by tribal groups. This tension was compounded in 1994 when the Indian Lands Open Dump Cleanup Act (25 USCA Section 3901-3911) was established in response to land and water contamination from unregulated and managed landfill sites. Given the resource limitations of the tribal communities the legislation meant that many open dumps had to be closed or handed over to local government or private companies. Despite this monitoring reports have found increased rather than decreased numbers of open dump sites and even the current figures are likely to be underestimating the problem because of a lack of access to remote areas (USHHS 1998). While Ortiz (2003) does find improvements within some tribal communities he concludes that the overall pattern of waste management across all tribal lands is highly uneven and dependent on the ability of individual tribes to obtain government resources. Successful activities include the appointment of tribal law enforcement officials with the power to confiscate vehicles involved in illegal dumping in Arizona and the development of a solid waste tribal code in collaboration with a private consultancy in Wisconsin, but both of these initiatives involve significant resource allocation that may not be available to all tribal communities. The problems of waste management in these contexts cannot though be reduced to financial constraints alone. There are on-going struggles over the sovereignty of tribes that authors such as Pardilla (1999) see as being rooted in a disregard by government agencies of the status and contribution that tribal communities can make to environmental policy making. Since the 1990s there have been a number of funding bodies and agencies developed to both build better relationships between state and tribal governments and upgrade waste management facilities appropriately which have produced good results but fail to deal with the scale of the problems identified by Ortiz (2003). He sees the development of partnerships between tribes and local governments or agencies as being a more productive means to improve waste management conditions and relations because they are entered into voluntarily and can be tailored to specific conditions in terms of the level of formality or informality of the partnership arrangement. The development of partnerships may provide channels for communication between tribal groups or between tribes, government bodies and non-tribal communities that can lead to opportunities for resource sharing. At the same time the level of expertise in technical and scientific knowledge amongst tribal members is increasing both within Indian country and within federal agencies such as the Environmental Protection Agency. Ortiz (2003) suggests that these new developments provide hope for more collaborative rather than confrontational relationships between tribes and government agencies but that ultimately success will depend up governments dealing sensitively with the cultural, historic and socio-economic conditions of those involved.

Asia: Collaboration and Constraint

Asia contains a diversity of waste management experiences as rich as any found elsewhere, although there is less available research specifically related to the waste governance field in this context than in either the UK or the USA.

China continues to make headlines in the western world for both its rapid economic rise and the potential impact that growing activity will have on economies and environments. Dong et al (2001) found that China already produces more than 29 per cent of the global municipal solid waste each year and if current economic trends continue this is likely to rise. At present however there is little accessible material on waste management practices and even less work on the challenges that lie ahead for that country. There are exceptions of course and the research of Yuan Hui et al (2006) examines the urban solid waste management practices in Chongqing. Yet even their analysis of challenges and opportunities for solid waste practices are divorced from in-depth cultural and political discussions common to waste governance analyses of other locations.

In other Asian countries a critical voice is easier to discern, particularly in the wealth of studies of waste management practices in India and Bangladesh (see Ahmed and Ali 2004; Agarwal et al. 2005; Forsyth 2005; Srivastava et al. 2005; Zurbrügg et al. 2004), but there are also examples from Sri Lanka (Vidanaarachchi et al. 2006) and the Philippines (Forsyth 2005). The clearest concern amongst this literature is the adoption of partnerships for waste management, particularly the patterns of power and control within those partnerships between communities and the private sector. As Ahmed and Ali (2004) note, both public and private sectors are active in managing waste in developing countries and there is an increasing trend to link them through Public Private Partnerships (PPP) with the aim of improving efficiency and creating new employment opportunities. However such partnerships have to consider the large number of people on the margins of such societies and how their needs – as both users and providers of waste services – might be served through partnership mechanisms.

Economic analyses of partnerships suggest that public-private associations can offer opportunities to exploit the advantages of both sectors, providing the means to combine 'the efficiency and expertise of the business world with public interest, accountability and broader planning of government' (Ahmed and Ali 2004, 467). Co-operative endeavours between the two sectors though may equally create a negative impact on those least able to cope with worsening life situations. In particular the consideration of people working within the 'informal private sector' is required. Ahmed and Ali (2004, 469) refer to this sector as 'initiatives utilising small amounts of capital and household or individual labour which operate outside government regulations for business activity'. Given the constraints under which local governments operate in many Asian countries, and the endemic persistence of poverty, unemployment and underemployment, the existence of a large informal sector is unsurprising. Chaterurvedi (1998) estimated that more than 150,000 waste pickers, as informal waste workers are often described, are active in the Delhi region alone, while Agarwal et al. (2005) found that the informal sector transports around 17 per cent of waste to recycling units in the same city. It is also widely recognized by these authors that workers in the informal sector are among the most vulnerable and marginalized in society. Waste pickers often use small scale waste buyers or recyclers to gain remuneration for their collected waste who may then pass on the material to larger scale operations. Informal waste actors are vulnerable both to changing conditions, as demonstrated by Snel's (1999) study of the impact a private sector company had on female waste pickers in Pune, India, and to exploitation (Begum 1999). While it has been suggested that the informal sector should be organized and brought into the formal realm because it has been shown to be an effective means of extending affordable services to poorer neighbourhoods, Agarwal et al. (2005) conclude that this process, if implemented in isolation, might actually lead to a dramatic rise in unemployment amongst the waste pickers.

The adoption of waste partnerships between the public and private sectors in Asia emerged primarily as a response to development aid from global economic institutions such as the World Bank and the International Monetary Fund and donor organizations such as USAID. As such it is argued that the resultant PPPs reflect the policy assumptions of these external organizations rather than responding to the indigenous context and are therefore particularly insensitive to the large workforce of the informal waste economy (Ahmed and Ali 2004). On the one hand there are conditions within Asian countries where PPPs might offer improvements to waste services. Where the public sector sees a large amount of its budget consumed by waste management or is generally unable to meet demands for services, a market opportunity for the private sector may exist as long as the public are willing (or able) to pay for improved collection and disposal conditions (Ali 1999). However there are also significant challenges for operationalizing PPPs. Many countries in Asia have weakly developed frameworks for regulating waste practices and certain private interests who benefit from current (poor) conditions of operation may seek to perpetuate the status quo rather than support increased regulation. In addition it is clear that systems of collection and disposal will need significant investment to cope with desired changes to the waste management system, not least because many people in developing Asian cities live in unplanned settlements with limited services. Such investment is not only financial in nature, but can also be allied to social investment in systems that foster transparency and accountability within waste partnerships to build trusting relationships across sectors and with communities. Therefore it is the precise nature of PPPs that will affect their appropriateness in resolving waste management issues in Asian countries as in other countries around the globe.

Tim Forsyth (2005) has conduced a critical examination of partnerships for waste management within an Asian context through an analysis of PPPs in waste-to-energy projects within the Philippines and India. He argues that there has been a tendency to see PPPs in these contexts as merely instrumental practices to collect and dispose of waste rather than as processes and places where systems of environmental governance can be formed. The key for Forsyth (2005) is to create the opportunity for deliberation within PPPs to facilitate the participation of publics in shaping social norms about environmental and waste issues that respond to their needs. Through such processes he suggests it might be possible to create responsive environmental policy and better relations between communities and waste management actors. The notion of broader partnership arrangements – including sub-state actors such as local government and sectors of civil society – has been promoted in global environmental governance institutions such as the United Nations and specifically at the World Summit on Sustainable Development held in 2002 (Plummer 2002). According to Forsyth (2005, 429) deliberative PPPs are defined as 'partnerships that maximize public debate about the purpose and inclusivity of collaboration between state, civil and market actors, as well as achieve the economic purposes of collaboration', but he warns that such expansive partnerships imply a greater sensitivity to the [re]creation of norms and values of the actors involved than has been the case previously. Forsyth is not the only commentator to warn of the difficulties of creating positive and sustainable partnerships (see also Osbourne 2000; Rosenau 2000) and he cites a range of concerns raised by other analysts including the power of vested business interests within a partnership to downgrade environmental concerns (Blowers 1998; Hajer 1995; Singleton 2000). Other authors comment on fears relating to the loss of democratic accountability when previously public services are transferred to a semi-private partnership arrangement that might serve to restrict the diversity of stakeholders who can participate in wider policy decisions about the best way to deal with waste matters (Rhodes 1996). Restricting the diversity of actors in partnerships risks developing institutional interactions that do not engage sufficiently with local contexts to ensure the development of locally appropriate mechanisms for managing waste. Nevertheless it has been argued that these concerns with accountability and local sensitivity can be placated whilst still achieving the benefits of private sector service provision by adopting co-operative and consensual decision making structures (Glasbergen, 1998; Meadowcraft, 1998).

As most of the advantages of co-operative mechanisms have been identified for western countries Forsyth (2005) considers a number of cases where collaborative relationships between developers and communities have been established within the waste-to-energy sector in an Asian context. From India he summarizes two contrasting cases – a biomethanation plant in Lucknow and a pyrolysis plant in Chennai – to examine whether the suggested benefits of partnerships emerge within different socio-political contexts. In Lucknow an Asian owned biomethanation plant was opened in 2003 to deal with municipal organic waste. The company works collaboratively with an NGO to support local waste pickers who are permitted to segregate waste before it arrives at the plant and remove non-organic recyclates in order to work within the current social norms of waste collection. The motivation for this co-operation was to protect the livelihoods of local urban poor, but it also ensured that the company retained a regular supply of organic material gathered at low cost. In contrast the Chennai case study NGO attempts to formalize the waste pickers activities were undermined by the local government's decision to replace public sector waste collection services with a private sector contract to a multinational waste company. This occurred at the same time that a pyrolysis plant was proposed by an international company. It was opposed by a national anti-incineration NGO because of fears of dioxin release and its potential impact on the livelihoods for waste pickers in the area. While the developers contested the negative environmental impact of the technology they, in contrast to the Lucknow case, challenged the prevailing social norm of waste picking, suggesting that it was unacceptable to allow people to work with raw materials in this way. Of course this expression of social concern is also convenient given that the technology requires the materials, such as paper, waste pickers would normally remove from the waste stream for recycling. To the dismay of opponents, and generating concerns over corruption, permission to operate was given by the local governor even after the Pollution Control Board for Chennai rejected the company's application in the first instance.

From the analysis of these case studies Forsyth (2005) concludes that, while participation and governance across Asia are not uniform processes, the benefits of collaborative partnerships as described by analysts studying cases in a western context are hard to replicate in the specific political context of India. He cites the restricted nature of political debates and the exclusion of certain social groups as reasons for the lack of political pluralism identified by Meadowcraft (1998) as a foundation for collaborative partnerships. Even when less powerful sections of society are involved in partnerships, such as in Lucknow, they are usually only permitted when more powerful agents also benefit from such an association. As Forsyth puts it 'poor sectors of society ... were often co-opted to support wider political arguments from more powerful actors, such as the adoption or rejection of a particular technology or institutional structure' (2005, 437). In both Chennai and Lucknow the private sector companies argued to include or exclude waste pickers for the same reason; that it was (in the eyes of the company) for the waste pickers own good. It is clear that in both cases the private sector defined the nature and form of the partnership between themselves and the waste pickers such that 'norms about environmental concern and accountability [were being] imposed by powerful actors both inside and beyond' the local area (Forsyth 2005, 438). Given this the level of deliberation in the associations was limited and new spaces for communication between investors and citizens were not being developed.

Although Forsyth (2005) demonstrated that, with respect to waste-to-energy schemes, collaborative practices between the private sector and communities were generally flawed Zürbrugg et al (2004) present a more optimistic case for such activities through decentralized composting initiatives for urban waste. Such activities generally develop as a necessary response by local communities to poor collection services from local governments and organic waste is collected in vacant plots and composted before being used or resold to the community as fertilizer for private gardens or parks. The composting initiatives require support within local communities to agree to have a composting facility in their locality and to participate financially to support the scheme as well as a level of organization in order to separate and collect the organic waste. For Zürbrugg et al (2004) the main challenge in such initiatives is not community support, but the actions of local authorities that can either serve to undermine or assist decentralized composting operations. In Pune and Mumbai authorities provide support mechanisms, through expertise and discussion fora, but in other cases such as Bangalore it is feared that the introduction of mandatory flat fee waste charges will undermine citizen commitment to composting initiatives. Here Zürbrugg et al (2004) call for negotiated agreements with neighbourhood composting initiatives to maintain the viability of such activities in association with the development of comprehensive municipal solid waste treatment strategies.

The need for negotiation and partnership in providing and operating waste management services in Asia is also a common theme in many studies of African waste management, the final area of illustrative case studies provide in this chapter.

Africa: Partnerships, Privatization and Politics

In 1997 Lusugga Kironde and Michael Yhdego examined solid waste management in urban areas of Tanzania using governance as a framing tool for their research. Through a case study of Dar es Salaam they identified governance in terms of central-local government relationship and the relationship between local government and international, national and community institutions and stakeholders. Although identifying a range of challenges for waste management as practiced in Dar es Salaam – including corruption, mistrust between politicians and publics and the politics of privatization – they did not rate a lack of financial resources as a major problem and advocated a partnership approach for public authorities and various stakeholders focused around a community-based strategy (Kironde and Yhdego 1997). Following this research Dar es Salaam has become a centre of waste governance studies in Africa. However others challenged the view that poor resources were not a limiting factor for waste management systems (Halla and Majani 1999). Despite this disagreement over the role of resources there remained a general commitment to promoting participatory and partnership arrangements for waste management in both policy and academic analyses (Onibokun and Kumuyi 1999). By the beginning of the twenty-first century Kaseva and Mbuligwe (2005) reported that the private sector had become the dominant force in solid waste management in the city. They suggested that the private sector had created a higher operating efficiency, employment and income from their waste activities because of their freedom from the bureaucratic hurdles that constrain public sector operators, citing improvements in waste collection from 10 per cent in 1994 to around 40 per cent in 2001 as evidence to back up their claims. Nevertheless problems still occurred in Dar es Salaam because of a lack of enforcement of waste legislation, illegal dumping, the non-payment of refuse collection charges and the large number of residents living in unplanned settlements that inhibits the collection of waste. Such problems are common to urban settlements across Tanzania including Zanzibar (Myers 2005) and have been reported in other African countries such as Kenya (Henry et al. 2006) and Zambia (Myers 2005).

While there is growing attention to governance issues throughout Africa the most in-depth analysis of waste governance comes from Garth Myers in his text *Disposable Cities: Garbage, Governance and Sustainable Development in Urban Africa* (2005). Myers uses questions about solid waste to examine issues of recent urbanization and development in Sub-Saharan Africa. As noted in the previous chapter he takes four themes – neoliberalism, sustainable development, good governance and the politics of cultural difference – to structure his study and as a result provides a critical framework for studying waste governance in Africa. He suggests that the privatization of economies and service delivery due to the neoliberalism demanded by donor agencies in Africa has led to a decline in government jobs in urban centres without a concomitant growth in employment through the private enterprises that have replaced them. Hansen and Vaa (2004) have linked this phenomenon to the rise in informal work practices and service delivery including waste management as people struggle to secure their livelihoods under conditions of scarce resources. The dominant influence of neoliberal agendas in development strategies have even been identified in the nature of sustainable development discourses being promoted in Africa through global institutions such as the United Nations Sustainable Cities Programme (UN SCP) (Hanson and Lake 2000). Myers, drawing on Abrahamsen (2000), argues that the interaction between neoliberal strategies and this particular understanding of sustainable development – focusing on development rather than livelihoods – has created a governance system that tends to exclude rather than empower local authorities and communities. This is despite the rhetoric of the Environmental Planning and Management (EPM) model that is central to the UN SCP. Such exclusion in conjunction with a legacy of colonial segregation based on race and class alongside contested gender and identity politics leads to a complex and potentially explosive cultural politics of waste in the region (Watts 2003). Most useful in this analysis is an attempt to simultaneously attend to cultural politics as well as environmental, economic and political conditions in the tradition of political ecology (see Blaikie and Brookfield 1987). Yet, wary of the charge of romanticising localities in political ecology analyses, Myers (2005) also identifies the situatedness of his research in global, national and local contexts.

In contrast to the more positive findings of earlier waste management analyses of Dar es Salaam, Myers (2005) contends that the rhetoric of the Sustainable Dar es

Salaam Programme (SDP), and particularly its plans for waste, conceals authoritarian rather than inclusive governance strategies that perpetuate colonial legacies. The cited improvements in collection and disposal (see Kaseva and Mbuligwe 2005), he suggests, are merely short term successes that came at the expense of reconstructing state-citizen relationships for long-term stability. The SDP began in 1989 when the Prime Minister accepted suggestions from UN-Habitat to adopt a new strategic planning framework that was being developed. In 1992 UN-Habitat provided finance for Dar es Salaam to promote a Sustainable Cities Programme. A city consultation was held with 350 participants from public, private and civil society from which improving solid waste management and providing services for currently unserviced settlements emerged as priorities. In 1996, after a short period of democratically elected local government, a City Commission was established to manage Dar es Salaam whose membership was entirely appointed by the Prime Minister. The Commission's Director implemented the privatization of solid waste management in the city and collection and disposal rates dramatically increased to an estimated 43 per cent in 2003 (Myers 2005, 45). In 1999 the City Commission gave way to another attempt to institutionalize democratically elected local government and Dar es Salaam was sub-divided into three municipalities (Iiala, Kinondoni and Temeke) that each had its own council and SDP office. The push towards further privatization of waste services continued despite this restructuring and despite the need for on-going donor assistance and recurrent organizational, political and management challenges (Kironde 2001). The endurance of privatization, according to Myers (2005), can be linked to the instrumental relationship between Tanzania and global economic players such as the World Bank, the International Monetary Fund and other aid donors during the 1990s and 2000s that demanded structural adjustment policies and free market principles. Tanzania's control over its path of development had effectively moved out of its own hands (Wange et al. 1998). The reforms demanded by structural adjustment impacted the ability of the public sector to provide services as they received less support from the central treasury at the same time as citizens ability to pay for services was being constrained by rising unemployment and prices increases.

The privatization programme began with a pilot project for ten of the inner city wards during 1994 when a new company Multinet was awarded a contract to collect waste from the central business district. The company invested 500,000 dollars of its own money in the venture and in 1996 the pilot was expanded into over half the wards of the city. By 2003 21 private companies and 23 civil society organizations collected waste in the city but only two of the larger private companies, including the original private actor Multinet, had more than four trucks for the collection of waste and they provided the bulk of waste collection (Myers 2005, 50). The city council and the SDP sought to facilitate a partnership between an organized union of scavengers and private firms in order to increase recycling. By 2003 Myers (2005, 53) reports that by 2003 nearly 75 per cent of glass, paper and plastic were separated at one dumpsite in the city elevating total recycling to between 10-15 per cent. In addition the continued existence of civil society organizations at least provided some employment opportunities for low-income residents as it was a requirement of contracts that workers were hired from within the communities that they served.

Although collection rates and volumes of waste being disposed of to landfill were increasing and some commentators praised the privatization for its efficiency and effectiveness (Majani 2002) there were still significant challenges relating to the collection of waste charges for services. In some places civil society organizations worked with private sector companies by bringing waste out from informal settlements to collection points where the private firms could then transport waste to disposal sites. In other areas civil society organizations were forced to close as contracts were removed. There were problems behind the scenes even where the UN-Habitat programme held up examples of civil society organizations as good practice leaders. Myers (2005) recounts how one organization, the Kinondoni Moscow Women's Development Association (Kimwoda), gained international praise for its work and was rewarded with a contract for waste collection from an entire ward where previously it had shared collections with other firms. However four other civil society organizations working in the same ward had had their community contracts eliminated to reward Kimwoda and this occurred at precisely the time when Kimwoda established itself as a chartered private company rather than a voluntary organization. This process of awarding or eliminating contracts lacked transparency and created problems between public, private and civil society sectors such that 'there was significant disregard for the negotiative, deliberative relationships [that were] supposed to be the bedrock of a new approach in the poor and rich communities of the city' (Myers 2005, 52).

Indeed while the development of partnerships between private companies and civil society organizations for solid waste management was seen as a central pillar of action plans within the SDP the overarching environmental aim was to remove the back-log of waste that had accumulated in the city and, at least initially, any means of achieving this was deemed acceptable. Beneath the commendable increase in collection rates however lies a more complex picture of a highly variable collection service and poor disposal practices. Olofsson and Sandow (2003) found that the poorer the area within the city the less likely it was to be served with a waste regular collection. Such patterns are unsurprising given the unwillingness of private companies to provide a service for which they would not be able to recoup costs, but it created a huge variation in service provision across the city that is not clear from raw statistics or necessarily visible to donors and wealthy elites. While waste collections were variable, disposal practices were more uniformly poor across the city. Myers (2005) found that under-resourced dumpsites created health, safety and environmental problems both for nearby communities and for employees who frequently worked without protective equipment. At these sites hazardous materials were mixed with municipal waste to the extent that both ground and surface water around dumps was polluted leading to high levels of toxicity in urban agricultural crops as a result of soil and waste contamination.

Concerns of the most marginalized about the side effects of poor waste management practices remain high but few channels exist for them to express their views formally through the SDP. Myers (2005) details the role of the SDP Director Keenja in promoting the liberalization of market forces rather than establishing the ideals of a more collaborative and co-operative planning style as advocated by the UN-Habitat Sustainable Cities Programme through its Environmental Planning and

Management (EPM) model. The lack of movement towards more inclusive planning cannot be blamed entirely on the state sector however and some commentators have concluded that while partnership discourses are prevalent amongst civil society organizations within Tanzania they have not been translated into actions because of vested interests and in-fighting between groups (Mercer 1999, 2003). While reviewers of the Sustainable Cities Programme have also claimed a lack of interest and participation amongst communities in Tanzania, the negative reactions of citizens are likely to be the result of a legitimate scepticism about the state's genuine commitment to partnership. Whatever the exact causes official channels of planning, be it for waste or other services, remain within the control of the state or private sector elites. Nonetheless it would be over-simplistic to represent citizens in Dar es Salaam as being without agency as the rise in unofficial activities has matched the neoliberal reforms. Kironde and Ngware (2000) and Halfani (1997) have all presented arguments that characterize citizens as actively creating informal systems of governance that run parallel to, but often contradict, governing agendas. Myers (2005) too recounts a series of examples of daily noncompliance by citizens as a means of survival within an unsympathetic neoliberal regime that fails to provide such basic services as waste management. He also identifies a growing number of associations, cooperatives and civil society organizations acting outside the SDP. Although most of the action reacts to crisis events rather than with a broad sense of forward planning, both donors and the state fail to give careful attention to the cultural politics of difference inherent in the civil society sector at their peril. For while much of the rhetoric of the SDP relies on the civil society organizations to help deliver action plans the majority of registered community groups in poor areas of Dar es Salaam are religiously affiliated and hostile to the government.

In summing up his view of waste governance in urban Africa Myers concludes that

> new forms [of governance] are rarely as progressive or liberatory as the rhetoric surrounding them. Neoliberal privatisation often seems to sow discord and selfishness; sustainable development programmes seldom improve either the environment or the livelihood of the poor; good governance recreates and improves upon the exclusionary democracies of late colonialism; and the politics of cultural difference produce debilitating battles over emplacing identity that usual leave the Other Sides of cities right where they were (2005, 15).

Conclusion

The purpose of this chapter was to illustrate the diversity of the geographical contexts in which waste management strategies operate and also to engage with a wide body of literature that has attempted to analyse waste management within those diverse contexts. In particular studies that have engaged with concepts of waste governance were identified and examined. The review inevitably only provides a snapshot of the wealth of information available on waste from around the globe. It is clearly an incomplete picture in that no case studies from Latin American or the states of the former Soviet Union are presented. In defence of the selection it was not the intention to describe as many waste management contexts as possible but rather to highlight a range of contrasting experiences where governments operating at different scales interact with public, private and civil society sectors. The chapter has shown that despite the, albeit restricted, diversity of the case studies there are a number of common themes that emerge in many locations around the world where societies struggle to deal with increasing volumes of waste within complex and often contested socio-economic and political environments. Inter-scalar interactions between governing bodies shape the landscape of waste management on which networks and particularly partnerships for waste management practices emerge. In turn partnerships and networks can impact on the nature of those governmental intersections in a dialectical encounter. It is in the detail of those associations that conflicts over waste become apparent and those conflicts can have social, political and environmental dimensions. Any understanding of waste governance therefore requires attention to the evolution of political structures, economic agendas and cultural conditions. In the next part of this book detailed attention is given to two countries that despite having certain commonalities in terms of contingent conditions have developed different discourses of waste management and strategies for waste governance.

PART 2
Governing Garbage: Case Studies

Chapter 4

A Comparative Framework: Contextual Background

Introduction

From the preceding chapters it is clear that in order to develop a comprehensive and geographically sensitive governance analysis of municipal waste attention must be paid to the nature of waste governing systems, that is their form and function, including consideration of participants, programmes and practices. However it is also important to contemplate why the waste governing systems operate the way they do, essentially reflecting on what forces promote or constrain certain interactions between tiers and spheres of governance. As waste practices do not operate within a vacuum this means that research endeavours have to transcend the waste sector and include wider socio-political, cultural and economic factors that might influence practices. Ultimately it is necessary to ascertain the outcomes that the governance systems generate.

Separating interventions, interactions and outcomes in this way is, of course, to some extent artificial. The reality of governing is dynamic and recursive with a shifting scene of actors that influence how interactions are played out and the shape of programmes that take precedence. Equally as particular programmes are operationalized so they can engender support or resistance from individuals and institutions who seek to influence outcomes through various interactive practices. Given the diversity of the waste arena these processes of participation and interaction can occur simultaneously, although perhaps in different ways, in relation to particular elements of waste management. At the same time it is necessary to provide some structure to the governance analysis and so, with the proviso that interventions, interactions and outcomes are intimately interrelated, a tripartite framework has been developed that considers policy interventions, interactions between actors in relation to those policy interventions and finally the outcomes of those interactions and interventions embedded within an appreciation of social and economic context. Aside from the pragmatism behind the tripartite framework, the structuring of the governance analysis in this way also facilitates comparative research. This is significant for there has been little concerted effort to compare and contrast waste governing systems in different contexts (although see Parto, 2005 for a useful initial foray in this area).

Ireland and New Zealand are examined in this volume in order to compare waste governing frameworks because, in spite of the vast geographical distance between them, there are a number of similarities in historical context and social structure between the two nations. Both have similar population sizes and distributions, they are both past colonial states with histories of agricultural dependency and both are strongly associated with environmental quality, particularly natural scenic beauty. Importantly for the concerns of this book, both countries have historically been dependent on landfill for the disposal of municipal waste, neither have municipal solid waste incineration and both are experiencing increases in the volumes of waste being produced. However such similarities sit alongside significant differences in contemporary political, economic and social conditions that affect the way that waste is conceptualized, managed and disposed of. This chapter will provide a brief overview of each case study country examining both economic and political structures and relationships between non-human nature and society as they have emerged in recent history. These considerations form the backdrop for a reflection on the environmental policy frameworks in which waste governing structures are situated, although not contained. The final section considers some of the methodological issues surrounding comparative governance analysis, the approach adopted and some inevitable limitations.

Ireland: an overview

> The sounds of Ireland, that restless whispering you never get away from, seeping out of low bushes and grass, heatherbells and fern, wrinkling bog pools, scraping tree branches, light hunting cloud, sound hounding sight, a hand ceaselessly combing and stroking the landscape, till the valley gleams like the pile upon a mountain pony's coat (Montague in Council of Europe 2005, 42).

Montague's *Windharp* remains an archetypal romantic vision of the Emerald Isle for many visitors to Ireland, emphasising its rurality, landscape and climate. Yet the complexity of Ireland's natures and cultures, and importantly their interaction, is not fully captured by such romanticism and it is necessary to look to wider discussions of political, economic and cultural development.

Political History and Economic Development

Ireland has been radically shaped, both physically and psychologically, by its political history under colonialism. It was part of the UK from the early 19th century until 1922 when the Irish Free State was formed, but it was not until 1949 that Ireland, or Éire, was declared a Republic. After decades of poor economic performance and mass emigration the tides of economic fortune changed during the 1990s and the Republic averaged 10 per cent growth in Gross Domestic Product (GDP) between 1995-2000. It was this period that gave Ireland the title of a Celtic Tiger economy (McAleese 2000). While growth rates in the economy dropped from the heady peak of the late 1990s following a global economic slow-down in 2005 they remain above the EU15 average (Department of Finance 2005). During this period agriculture, once the most productive sector for the economy and a powerful lobby group in government, became dwarfed by business and industrial developments with the agri-food sector only accounting for 8 per cent of GDP and around 25 per cent of net foreign earnings (Teagasc 2005). In addition to revenue from exports, consumer spending, high levels of construction and continued business investment have bolstered the strength of Ireland's economy.

The extent of the economic boom experienced during the latter half of the 1990s was all the more surprising to outside observers because of the small size of the country with just over 7 million hectares of land (or 70,000km^2) and a population of just over 4 million people (CSO 2006, 9). The country is sparsely populated, around 61 people per square kilometre (United Nations, 2005) with only a few, relatively small, urban centres. The 2006 census records Dublin – the capital – as the largest city with just over 500,000 inhabitants, followed by Cork (119,113), Galway (71,983) Limerick (52,560), and Waterford (45,775) (CSO 2006, 21). However the population grew by over a million from 1971–2005 boosted by returning nationals and increasing numbers of immigrants attracted to the buoyant economy. Geographically located in close proximity to the UK and just off mainland Europe, Ireland joined the European Economic Community (now the European Union) in 1973 and became part of the Euro zone in 2001. As such the EU has been a pivotal feature of Ireland's development. Substantial funds for structural development came from the EU giving an added impetus to investment and the requirements of European Directives and Treaties have actively shaped to domestic policies, particularly in the environmental arena.

Politically Ireland is a parliamentary democracy. The Oireachtas is the national parliament and it consists of the President and two houses. The two houses are the Dáil (The House of Representatives) and the Seanad (the Senate) and they take their respective powers from the Constitution of Ireland and law (Dooney and O'Toole 1998), however the government is responsible to the Dáil only, thus making it the primary locus of government. The Constitution of Ireland, Bunreacht na hÉireann, was adopted by referendum in 1937 and it defines Ireland as a sovereign, independent and democratic state. It sets out the administrative structure of the Government as well as the principles of legal and social policy to guide the Oireachtas. The President of Ireland is elected by direct vote from the people for a term of seven years. While the position is primarily a ceremonial one the President is essentially the guardian of the Constitution and may choose to exercise these powers on the advice of the Government or Council of State. The Head of the Government is the Taoiseach who is appointed by the President on the nomination of the Dáil, while civil servants assist in the running of each of the fifteen Departments of State and are appointed through public competition (Government of Ireland 2005). Voting in general elections is by a system of proportional representation and a single transferable vote in multi-seat constituencies (Oasis 2005).

Sub-national government in Ireland is relatively weak when compared to many of its European counterparts although an amendment to the Constitution of Ireland in 1999 gave clear constitutional status to local government for the first time and made it a mandatory requirement for local elections to be held every five years (Callanan and Keogan 2003). Local government in Ireland is made up of 29 county councils, five city councils and 75 town councils. There is at least one council for each county, Dublin has three (South Dublin, Dun Laoghaire-Rathdown and Fingal) and Tipperary has two (North and South). The location of these counties and cities is presented in Figure 4.1. Each county has elected councillors, with the number being defined according to the population size, and a chief executive called a county or city manager is appointed by central government to oversee the management of the local authorities.[1] Local authorities in Ireland are responsible for the provision of a range of public services including housing, planning, roads, water supply and sewerage, development incentives and controls, elements of environmental protection, recreation facilities and amenities, and agriculture, education, health and welfare. They are also supposed to promote the interests of the local community through social, economic, environmental, recreational, cultural, community or general development. These functions of local authorities are carried out through different mechanisms, some are enacted by the members of the authority acting as a body at meetings, some are carried out by committees and some are the responsibility of the county or city manager.

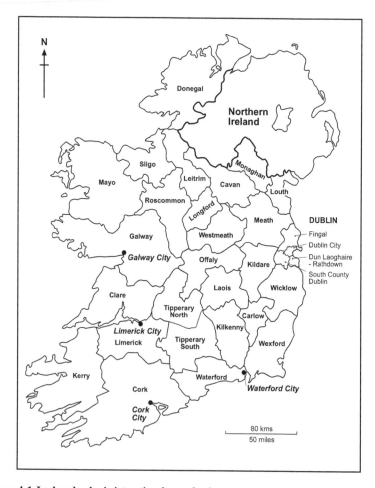

Figure 4.1 Ireland administrative boundaries

---

1   Although a few counties share a manager such as Leitrim and Sligo.

Local government funding in Ireland is currently reliant on income from a combination of commercial rates, charges for goods and services and transfers from central government through the central block grant. Only 56 per cent of funding is generated locally leading to concerns about limited discretion for local authorities, a lack of revenue and inequitable funding systems (Indecon 2005, i). Critics of the centralized funding system call for an increase in locally based charges for services provided by local authorities, such as waste management, in order to 'bring many benefits such as efficiency, transparency and consistency with the polluter pays principle' (Indecon 2005, x). Any such changes to more locally raised charges would however require careful construction in order to generate public acceptability as recent protests against local waste charges in Dublin have illustrated (Davies 2007).

Even more so than local government, regional government has been a weak legislative force within Ireland's politically centralized administration. While there are eight regional authorities and two regional assemblies their roles are limited and their members are nominated by local authorities rather than elected by the general public. The regional authorities co-ordinate certain activities of the local authorities and monitor the use of EU Structural Funds while the regional assemblies manage the regional programme of the National Development Plan. The regional authorities were established by the 1991 Local Government Act and came into existence in 1994. They had the specific responsibility of reviewing the Development Plans of local authorities in their region and in adjoining regions, preparing Regional Planning Guidelines and Regional Economic and Social Strategies and promoting cooperation, joint actions, arrangements and consultation among local authorities and other public bodies. Each regional authority, financed largely by the constituent local authorities, has an operational committee, an EU operational committee, a designated city/county manager from one of its local authorities, and a Director (assisted by a number of policy and administrative staff). Two regional assemblies, the Border, Midlands and Western Assembly and the South and Eastern Assembly, emerged in 1999 following negotiations by the Irish Government with the EU in relation to the management of Structural Funds. These groupings are responsible for managing the Regional Operational Programmes under the National Development Plan, monitoring the general impact of all EU programmes under the National Development Plan/Community Support Framework and promoting coordination in the provision of public services in the Assembly areas. Waste management regions have also emerged, and these will be addressed in detail in later sections of this chapter, but these do not map onto either of the pre-existing regional structures and their remit is restricted to waste matters.

While the above suggests there is a clear centralist strategy within Irish political structures there has also been a tradition of partnership between government, industry and social actors through the negotiation of Social Partnership Agreements (SPA) since 1987. Social partnership describes an approach to governing where interest groups outside elected representatives have an opportunity to contribute to decision making. The agreements focus principally on incomes, fiscal, social, economic and competitiveness policies and are negotiated between the Government and the social partners who are currently categorized into four pillars: trade unions; employers and business; farming; and the community and voluntary sector.[2] In addition to national level partnership through the SPA there have also been local initiatives to try and develop more inclusive and deliberative practices in areas such as community development and planning. This has particularly manifested itself through institutional reform at the local level and the formation of city/county development boards (CDB) and strategic policy committees. The CDBs, developed in 2000, are led by local government, but include representatives of local development bodies together with the state agencies and social partners operating locally. For the first time CDBs brought together key players at the local level to engage in a process of long-term planning for each county or city. Taylor (2001) suggests that there are tensions between the agreement negotiation process, which was initially at least envisaged as a social democratic project, and its products that contain elements of a neo-liberal economic and political project. However the impact of this broad shift towards consultation and participation indicated by partnership agreements, local policy committees and development boards has yet to be critically examined in any depth. An area that has achieved more attention is the interrelationship between nature, culture and society in Ireland and some of the key debates are considered in the following section.

Nature, Culture and Society

> It is often called the "Emerald Isle", the land of "forty shades of green". An island surrounded by the North Atlantic Ocean and the Irish Sea, it is a land of peat bogs, coastal marshes, high cliffs, soft green pastures defined by near stone walls, and fields ablaze with wildflowers. It is a lush place, where clear skies repeatedly yield to gentle rain. It is Ireland (Kricher, in Viney 2003, vii).

Interactions between society and nature have impacted the landscape of Ireland since the Neolithic farmers of some 6000 years ago settled, cleared woodlands and grew crops, but it was with population expansion in the late 18th century that landscape and wildlife resources were first put under severe pressure. By 1841 Ireland was home to over 8 million people; nearly double the current population. This was fuelled by the cultivation of the potato that enabled large swathes of previously marginal land to be brought into productive use. The expanded population had a negative impact on the ecology of the countryside, particularly in terms of trees and woody shrubs that were used for fuel (Mitchell and Ryan, 1997), but it is the famine of 1845 that remains the most significant reference point for Irish conceptions of human environment relations. The devastating implications of the potato famine on Irish society are well documented (see for example Donnelly 2000; Howe 2000; Ó Gráda 2004), but it also played a role in shaping attitudes towards the environment in two ways. First, despite its social and political roots, the famine created a sense of betrayal by nature and second, it generated a sceptical view of state-based interventions in environmental management (Foster 1997). Unsurprisingly perhaps 'in the impoverished early decades of an independent Ireland, the popular view of

---

2   The community and voluntary pillar was formally established in 1996 and included in the Partnership 2000 agreement (Larragy 2006).

nature was urgently utilitarian and land-hungry' (Viney 2003, 1), but it has also been suggested that environmental matters were seen to be the concern of privileged elites so that environmental practices such as nature conservation were regarded as alien to everyday life in Ireland (Cabot 1999). In contrast to the romantic poets of England, such as Keats and Wordsworth, Foster (1997) reports that there was little celebration of nature for its own sake amongst their Irish counterparts. Indeed he suggests that 'the words "nature", "landscape" and "scenery" ... have among the bulk of the Irish people to this day a somewhat effete connotation and evoke an Anglo-Irish world view' (1997, 412). The little deliberation on nature that does exist in the works of Irish romantic artists and poets focused on the picturesque ruination of the landscape under colonialism and highlighted the vestiges of pre-colonial dignity. Such readings allied nationalism to unspoilt nature with the rallying call that 'nature would resurge to overwhelm the superficial cultivation of the established order' (Foster 1997, 415). While there were concerns expressed about the impact of development on historical, cultural and physical environments during the 1960s attention to environmental protection was less well developed in Ireland than in other European countries. In part this was because of weaker development pressure on the land, but it was also because the organizations so important in shaping public opinion towards conservation, such as An Taisce, were still tainted by their association with privileged elites. Indeed Feehan remarks that the identification of nature conservation, and by association environmental concern, as a recreational activity of privileged elites survived well beyond its origins in the romantic period of the late 19th and early 20th centuries to become 'one of the most stubborn of all obstacles in the campaign to educate the community to an environmental consciousness' (1997, 583).

Examination of the attitudes and actions of the Irish public towards the environment has become commonplace in recent periods and it was in 1993 that Ireland first participated in the International Social Survey Programme that contained an environmental module (Faughnan and McCabe 1998). Respondents were asked about their attitudes to nature, science and the economy, their concerns about environmental issues and their sense of responsibility in terms of dealing with those concerns. They were also asked to detail their level of participation in environmental activities and provide their perspectives on various mechanisms for environmental protection. The overarching conclusion was that in comparison to their European counterparts Irish respondents were more concerned about economic development than environmental protection. While environmental quality, particularly local environmental quality, was identified as problematic more than half of the respondents felt that 'people worry too much about the environment and not enough about prices and jobs' (Faughan and McCabe 1998, 61). Such findings were unsurprising given the decades of poor economic performance and associated high unemployment and emigration of previous decades, but an explanation for the preoccupation with local environmental quality over more global environmental concerns is not so easy to discern. An answer is partially provided by Leonard (2006) who, in his analysis of the Irish environmental movement since the 1960s, suggests that recent social history has been characterized by community challenges to multinational developments or infrastructural projects formulated from a populist rural sentiment and a localized sense of place. Leonard sees these senses and sentiments as being relics from past periods when Ireland was primarily an agrarian and rural society. Yet as detailed in the previous section the economic landscape for most Irish residents has altered dramatically since 1993 bringing with it associated social, material and political changes. It would be expected that these changes – often characterized as a period of modernization – would also affect the discourses of governing institutions and governed entities in terms of how environmental issues are conceived and concerns articulated. Growth in wealth has certainly contributed to greater consumer spending and consumption of goods, services and resources. By association this has led to increasing by-products of consumption such as global greenhouse gases, air pollution and particularly waste. As consumer confidence grew there were concerns expressed that the traditional pillars of Irish society, the church and the family, would come under pressure. On the one hand high profile cases of corruption and abuse in both the church and politics have fostered scepticism about public figures and institutional procedures generally. On the other the number of households has increased through inward migration and decreasing household size leading to greater demands on land, energy and materials and to debates about an atomization of social structures and a decline in the influence of the extended family (McDonald 2006).

At the same time Ireland has become a more urban society with increased suburbanization of cities and towns as agricultural opportunities declined and rural to urban migration occurred. This shift has contributed to changes in the physical landscape of many areas by increasing pressure for new road developments and housing estates. The pressure for development is not, however, all one way traffic into the city and there are increasingly heated discussions about the desirability and sustainability of housing developments in the countryside (McDonald and Nix 2005). In general however environmental topics, such as housing, flooding and waste, have become a familiar feature within popular media and sustainable development discourses are now more frequently articulated in policy circles. The interpretation of sustainable development in the Irish policy context promotes the view that environmental and economic objectives can be attained in parallel. The Irish state has thus been described as ecologically modern in its outlook (Pepper 1999; Taylor 2001). Environmental regulation and control of pollution is perceived as providing a stimulus for technical innovation for cleaner technologies and eradicating the inefficiencies of pollution through the polluter pays principle. The push towards ecological modernization in Ireland can be linked to pressure through EU Directives to implement such mechanisms as Environmental Impact Assessment (EIA), Integrated Pollution Prevention Control (IPPC) and the Rural Environmental Protection Scheme (REPS), although it is also an interpretation of sustainability that lends itself more easily to the development project of successive governments in recent decades. Following in this vein there are concerns that underneath the win-win rhetoric of ecological modernization lies an ever powerful economic imperative (Taylor 2001). Expressing similar concerns MacDonald and Nix ask

> do we have any idea where we are going, any idea at all about the kind of Ireland being created during these years of prosperity? Do we care about the indelible imprint we're making on the landscape and the woeful legacy of 'development' we're leaving for future generations to clean up – if they can? (2006, 33).

At the same time the environmental movement in Ireland has yet to establish a strong national presence in policy making or generate a mass support base amongst the general public outside community based collective actions against site specific developments or pollution incidents (Allen 2004; Leonard 2006). As Viney (2003, 307-8) concludes 'environmentalists are still widely regarded as interfering, city-based do-gooders, and nature conservation is still largely identified with an Anglo Irish culture'.

Despite the social and economic changes that occurred in Ireland during the 1990s a comparison of environmental attitudes between the survey conducted in 1993 and a subsequent one undertaken in 2002 reveals relatively minor shifts in patterns of responses (Motherway et al. 2003). For example while overall levels of environmental concern remained constant across the period more people accepted that their actions can make a difference to environmental quality while fewer (although still a majority) claim to do what is right for the environment when this places more pressure on time or resources. In 1993 the major divergence of Irish environmental attitudes from European averages had been the low priority given to global environmental concerns, but in 2002 more people expressed concern about global air pollution and climate change suggesting increased awareness of global environmental issues. Nevertheless while environmentalism in 21st century Ireland appears to be emerging as a more mainstream concern pro-environmental practices and environmental activism are still far less common than in Scandinavian countries, Germany or the Netherlands (Kelly et al. 2003a; 2003b). So while the value of Ireland's green image, the Emerald Isle, is much traded on in tourist literature it is less clear whether such rhetoric moves beyond a descriptive moniker given to a land undoubtedly verdant as a result of its particular climatic conditions. This issue of greenness is revisited in the following section that maps out the landscape of environmental policy evolution in Ireland.

Environmental Policy

The nature of waste governance and particularly the evolution of waste policy in Ireland owes much to wider changes in environmental and local government policy since the 1970s and in particular to the growing influence of European legislation on those policy areas. Environmental policy making in Ireland during the 1970s and 1980s was characterized by an institutional split between the Department of the Environment, which was formed in 1977 to develop policies, and local authorities that were charged with implementing the resulting legislation. Issues of air and water pollution, waste management and sanitation were, through a series of Local Government Acts, made the responsibility of local authorities and this contributed to the sense of environmental matters being seen as local political issues. As in the UK, the model for environmental legislation in Ireland followed a path of incrementalism whereby new elements were added on to existing structures and policy responses tended to be ad hoc and reactive to external conditions (for example increasing European pressure) rather than comprehensive, proactive or integrated measures for environmental protection. In contrast to the rule making structures of environmental policy that emerged in Germany and the USA, Ireland opted for a more discretionary system based on voluntary codes of practice for regulated activities. In addition the environmental policies of the period avoided setting precise standards or defining principles and were developed in an atmosphere of negotiated compliance between regulators and business interests, which has been described by some commentators as indicative of a clientilist politics (Higgins 1982). The process of negotiated compliance in Irish environmental politics created an uneven playing field for interest groups. In the case of the EU Habitats Directive, for example, the Irish Peatland Conservation Council (IPCC) was only allowed one meeting with the Department of Arts, Culture and the Gaeltacht in which they were shown a draft of the regulations and that was presented as a fait accompli. On the other hand the agricultural lobby were involved in two days of talks around the issue of compensation. As Taylor succinctly puts it 'it is one thing to be allowed to enter the office, quite another to be allowed to stay and influence decisions' (2001, 14).

The ad hoc approach to environmental legislation and the reliance on negotiation for regulation was particularly striking in the waste management field. Local authorities were only responsible for disposing of domestic waste yet in practice all kinds of waste from both private and public waste collectors arrived at local authority dump sites and although local authorities were identified as the responsible regulatory authority for waste no enforcement officers were employed during this period and few prosecutions were therefore made (Scannell 1990). There was then a discrepancy between the defined legislative intent of policy and the practical implementation of that policy; an implementation deficit. During this period there was a considerable degree of pragmatism behind the regulatory approach which was heavily grounded in the environmental regulators having to pay heed to the economic repercussions of imposing stringent environmental regulations that might negatively impact on already scarce industrial activity (Scannell 1982). Effectively local authorities were in competition with each other as development corporations while also being exempt from many of the pollution controls they were supposed to be enforcing on others. Concerns that more stringent enforcement of environmental regulations might affect Ireland's attractiveness for inward investment from multinationals remained high (Leonard 1988; Scannell 1990). As Taylor (2001) points out impacts on the environment resulting from industrial development was only one of many competing concerns during this formative period of environmental policy and organizations such as Ireland's Industrial Development Authority (IDA) and the Institute for Industrial Research and Standards (IIRS) were pivotal in providing counter positions to an increasingly vocal environmental lobby. There was obvious tension between the dual roles of local authorities as both promoter and regulator of industrial activities.

At the heart of the difficulties facing environmental policy makers in the 1980s was the lack of resources allocated to the local authorities that had significant responsibilities for the implementation and enforcement of environmental regulations. This position was exacerbated when the number of Directives emanating from the EU began to grow significantly. Not only were time and personnel a problem, there was also a paucity of expertise within the local authorities to implement effectively the legislation that was in place and issues arose when transposing Directives into Irish policy. A key problem was that, despite the flexibility to allow national regimes to implement policies that are sensitive to national regulatory styles inherent in the Directives, there was a tendency to simply translate them verbatim into Irish law (Coyle 1994).

While representatives from environmental lobbying organizations were not powerful actors in industrial and agricultural decision making during the 1970s and 1980s they were, with the help of active community campaigns, able to bring environmental issues to popular attention and towards the end of the 1980s cross party support for a new approach to environmental policy regulation was being crystallized. The new approach to environmental policy regulation, mooted by the ruling coalition in 1989, was to be characterized by the development of a single agency – the Environmental Protection Agency (EPA) – to overcome the limitations of the previous ad hoc, reactive and clientilist regime. The Industrial Policy Review Group (1992) reported that this proposal was the institutional manifestation of a newly found confidence to protect the environment while also participating in the global economy; a particular form of ecological modernization where both economy and environment can be developed in a positive sum game (Pepper 1999). The resulting 1992 Environmental Protection Agency Act did indeed contain elements associated with the various interpretations of ecological modernization that included Integrated Pollution Control (IPC) licensing, a refocusing of attention on prevention rather than abatement and greater co-ordination and transparency in practices (see Christoff 1996; Hajer 1995; Weale 1992). The EPA was given a range of statutory functions including IPC licensing, but it was also required to provide a range of support services for local authorities, conduct state of the environment reporting and environmental research. Most significantly however it was the enforcement powers of the EPA and the ability to impose significant fines for noncompliance that were seen as the most progressive elements of the legislation. Nonetheless there remained concern that despite facilitating improvements the 1992 EPA Act permitted the continuation of what Taylor calls 'the Irish style of policy making ... [a] soft regulatory ethos' (2001, 42). For example the EPA was still reliant on self-monitoring and voluntary compliance by companies and local authorities in terms of reporting the impacts of environmental activities. There also remains some confusion regarding the relative roles of the EPA and planning authorities (and An Bord Pleanála, the Irish planning Appeals Board) in terms of environmental protection.[3] Before the formation of the EPA local authorities attached conditions to planning permission in an attempt to control pollution. The responsibility for pollution control through IPC licensing now lies with the EPA although planning authorities are still responsible for other considerations such as visual impacts, traffic and landscape. This may appear to be a clear delineation, but Taylor (2001) argues that complications arise because planning authorities or An Bord Pleanála are not able to consider issues relating to potential environmental pollution from the operation of a project where an IPC licence is

---

3 Applications for planning permission are dealt with initially by local planning authorities however if any decision (or conditions attached to that decision) is appealed then the central independent third party appeals system is brought into play. The system is operated by An Bord Pleanála the planning appeals board.

required (e.g. a waste incinerator) although they may consider pollution generated during the development of the project.

Essentially the planning system and the EPA are processes operating in parallel and with the enactment of the 2006 the Planning and Development (Strategic Infrastructure) Act there is a clear move towards centralising the planning system to mirror EPA activities. The Act introduced a consent process for infrastructural developments, provided either by statutory bodies and private developers, considered to be of national importance. Although not directly specified in the Act major infrastructural developments could include large scale waste facilities such as incinerators or regional landfills. The Act, which came into force on 31st January 2007, included the restructuring of An Bord Pleanála to allow for the creation of a specific division within the organization to deal with all major infrastructure projects. The Act also institutionalizes the requirement for developers to provide direct and substantial benefits to local communities affected by major infrastructural projects as well as opening up the possibility of pre-application discussions with applicants in relation to infrastructure consent. Essentially the Act requires the planning appeals board to have regard to 'the national interest, any effect the decision may have on issues of strategic economic or social importance to the State' (DoEHLG 2006a, 56).

It is not only in relation to planning issues that the waters of responsibility for environmental protection get muddied. Certain agricultural activities and their by-products (for example slurry or farm sludge) for example, are exempt from IPC licensing. Rather than indicating a new approach to regulation such exemptions suggested a perpetuation of past practices and patterns of power and influence in environmental policy communities.

Despite the gains made through the EPA in terms of introducing improved monitoring techniques and providing environmental information, rates of non-compliance remain high. Rarely however has the agency's ability to impose significant fines been invoked. There is a general perception that the EPA prefers to accommodate rather than punish wayward polluters and processes permitting objections to licenses from third parties are becoming increasingly restrictive. For although the EPA Act incorporated a role for oral hearings in the case of objections any decision to hold one remains at the discretion of the EPA. The hearing would be conducted by a person appointed by the EPA and would take place within a specified (and limited) time frame. Bearing in mind that the burden of proof in terms of identifying environmental impact lies squarely on the shoulders of third party objectors such constraints provide considerable challenges to all but the most organized and well-supported of environmental groups. Oral hearings then do not occur simply because of a large number of objections have been made but only when the EPA deem that the objections have a scientifically valid objection to the proposed development. In contrast the EPA has made clear assurances that those applying for EPA licenses will receive considerable assistance and pre-application clarification of EPA requirements, leaving Taylor (2001) to suggest that

> [the] environmental policy debate in Ireland is concerned no longer with the extent of ecological degradation, the quality of the environment or encouraging environmental

sensitivity, but with the complicated process of organising consent around new definitions of the extent to which pollution can be justified (Taylor, 2001, 5).

Such a system raises considerable questions about the role of science and public participation in environmental issues. These questions are very much to do with the style of environmental governance. Essentially the evolution of environmental policy in Ireland has emerged out of inherently political developments occurring at a range of different scales, from the European to the local, and through the particular interactions between governments, lobby groups and communities that are couched in historically generated and surprisingly persistent patterns of access and influence. So how does this compare with the experience of New Zealand? The following section reflects on New Zealand's economic and political histories and structures as well as interactions between people and non-human nature before examining the evolution of environmental policy.

New Zealand

> Imagine that you live in Asia, or Britain, or perhaps the US. You have driven home through the smog to your cramped apartment, and as you eat your dinner you see on TV images of snow-capped mountains reflected in crystal-clear unpolluted lakes. Cows graze in lush green pastures, native birds sing in the forests, waves thunder onto deserted beaches, and happy healthy people are having fun. It is New Zealand, and it looks like paradise (Ministry for the Environment 2001, 1).

Although the idea of New Zealand as a haven of clean and green environments permeates both the collective psyche of New Zealanders and many travellers to the islands, the claim to paradise articulated by the Minister for the Environment above requires careful analysis of political and economic developments as well as the interactions of nature, culture and society.

Political and Economic Development

New Zealand, or Aotearoa, is comprised of two main islands (North and South) that are more than 2000km away from their nearest neighbour Australia. The two islands, that are similar in land mass to Scotland and England (Gunder and Mouat 2002), accommodate approximately 4 million people. As a result the country remains sparsely populated, at around 15 people per square kilometre (United Nations 2005), with a few urban centres. Over half the population lives in the urban areas of Auckland, Hamilton, Wellington and Christchurch with one in three of the population living in Auckland alone (Te Tari Tatau 2005). Overall the population is aging and becoming more ethnically diverse with increasing numbers of Asian and pacific peoples moving to the country, but current levels of migration remain low and below replacement fertility rates mean that the population grew by just 0.8 per cent between 2000 and 2005 and is unlikely to rise above five million in the next 30 years.

Politically New Zealand is an independent state: a monarchy with a parliamentary government. Queen Elizabeth II (of the United Kingdom) has the title Queen of New Zealand. New Zealand's constitutional history can be traced back to 1840 when, through the Treaty of Waitangi, the Māori people exchanged their sovereignty for the guarantees of the treaty and New Zealand became a British colony. Although as detailed below, the interpretation of this treaty remains contested. The Governor-General is the representative of the Sovereign in New Zealand. New Zealand government has three branches: the Legislature, the Executive and the Judiciary, and power is divided between these branches preventing any one from acting against the basic constitutional principles of the country. Although each branch has a different role, they are not totally separate from each other. New Zealand has a single chamber of Parliament known as the House of Representatives that enacts laws, provides a government, supervises the government's administration, allocates funding for government agencies and services as well as providing redress for grievances by way of petition.

Following a national referendum in 1993 the parliament is elected using the mixed member proportional (MMP) system. Under the MMP system voters have two votes; a party vote and an electorate vote. Voters can choose what party they want in Parliament with their party vote and which person they want to represent their electorate with their electorate vote. New Zealand is divided geographically into 61 general electorates and six Māori ones. There are also 53 seats for list MPs. People of Māori descent can choose whether to be on the Māori or general electoral rolls with the number of seats changing according to the number of voters on the Māori roll. The party, or coalition of parties, that can command a majority of the votes in the House of Representatives forms the Government. The House's responsibilities are to debate and pass legislation, provide a Government, supervise the Government's administration by requiring it to explain policies and actions, supply money, and represent the views of the people of New Zealand. It has a number of Select Committees which examine proposed legislation (Bills) in detail, often hearing submissions from interested members of the public.

New Zealand does not have a single written constitution instead its constitutional arrangements can be found in a number of key documents, which together with New Zealand's constitutional conventions, form the nation's constitution. Key written sources include the Constitution Act 1986, the New Zealand Bill Of Rights Act 1990, the Electoral Act 1993, the Treaty of Waitangi and the Standing Orders of the House of Representatives. The Treaty of Waitangi was signed in 1840, as an agreement between the British Crown and a large number of the Māori of New Zealand. Today the Treaty is widely accepted to be a constitutional document, which establishes and guides relationships between the Crown in New Zealand (as embodied by the government) and Māori. The Treaty of Waitangi had at its heart a promise to protect a living Māori culture; to enable Māori to continue to live in New Zealand as Māori while at the same time conferring on the Crown the right to govern in the interests of all New Zealanders. This means that relationships between the Government and Māori in terms of defining and protecting cultural identity within a wider societal context are ongoing.

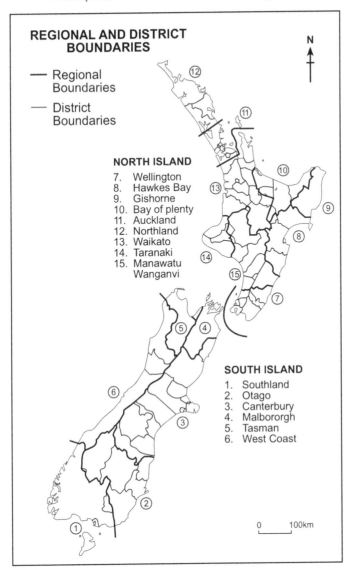

REGIONAL AND DISTRICT
BOUNDARIES

— Regional
Boundaries

— District
Boundaries

N

NORTH ISLAND
7. Wellington
8. Hawkes Bay
9. Gishorne
10. Bay of plenty
11. Auckland
12. Northland
13. Waikato
14. Taranaki
15. Manawatu
    Wanganvi

SOUTH ISLAND
1. Southland
2. Otago
3. Canterbury
4. Malbororgh
5. Tasman
6. West Coast

0   100km

Figure 4.2 New Zealand administrative boundaries

Local government, the lead local government agency representing all councils in New Zealand, is comprised of 12 regional councils and 74 territorial authorities (see Figure 4.2). Of the 74 territorial authorities 16 are city councils and 58 are district councils. Gisborne, Malborough, Tasman, Nelson City and the Chatham Islands are unitary authorities that combine regional and territorial council functions. These bodies are creatures of statute, but they are also autonomous and accountable to communities. New Zealand is a unitary state as opposed to a federation such that the authority of the central government creates regions. As a result local government in

New Zealand has only the powers conferred upon it by Parliament. These powers have traditionally been distinctly fewer than in some other countries. For example, police and education are run by central government, while the provision of low-cost housing is optional for local councils. While local authorities used to have control over ports as well as gas and electricity supply nearly all these functions were privatized during the 1980s and 1990s.

Regional authorities have primary responsibility for environmental management, including water, contaminant discharge and coastal management, river and lake management including flood and drainage control, regional land management; regional transport (including public transport), biosecurity or pest management. Territorial authorities are responsible for: local-level land use management (urban and rural planning); network utility services such as water, sewerage, stormwater and solid waste management; local roads; libraries; parks and reserves; and community development. Property rates (land taxes) are used to fund both regional and territorial government activities. In practice there is often significant co-operation on issues between regional and territorial councils as their roles frequently overlap. For example, in the waste field an application to build or extend a landfill would require consent from both the regional and territorial authorities as it has implications for land use planning and pollution control. There are 1098 elected councillors in local government (and 796 community board members) who are citizens elected on the basis of their understanding of, and potential contribution to, community issues. They are free agents elected to develop policy and long-term direction for the community of interest they represent. These local governors are directly accountable to, and are representative of, their communities.

New Zealand was, for a long time, characterized as a rural economy with sheep raising, dairying and beef production providing the foundations for employment and exports. Indeed more than half of the nation's exports were animal products of one kind or another during the 1990s and a much quoted statistic was that there were 20 times more sheep than people on the islands (McKnight 1995, 156). Given the small population New Zealand has always depended on foreign trade for its development. In the past, and unsurprising given the colonial links, the major trading partner in terms of exporting animal products and importing manufactured jobs was the UK. However when the UK joined the European Economic Community New Zealand's favoured trade position was affected. This change was a significant blow to the economy stimulating a period of restructuring and a search for new markets that included sweeping neo-liberal reforms across social, economic and administrative arenas during the 1980s. The aim of the changes was to move away from the heavily regulated and protectionist approaches that dominated New Zealand's trading systems and align them with the growing internationalism of global capitalism (Britton et al. 2002). At the same time the strength of global environmental governance was increasing and sustainability was becoming the common currency of international environmental regimes. In 1981 the Organization for Economic Cooperation and Development conducted a review of New Zealand's environmental administration and drew attention to its limitations. This, combined with increasing public concern for environmental quality, led to a reorganization and strengthening of environmental management creating what has been described as an 'uneasy mix of free marketeering and government intervention' (Cocklin and Furuseth 1994, 459). Key reforms to environmental management are addressed later, but they included the reorganization of central government agencies responsible for environmental management and resource development as well as a restructuring of sub-national government and legislative changes that culminated in the formation of the 1991 Resource Management Act.

Since the mid 1990s Asian markets have emerged as major zones of trade for New Zealand although the USA and the Middle East now also play an important role. Nonetheless economists continue to be concerned about the reliance on exports of primary products, particularly agricultural products, in order to be able to support the high demand for imports. As a result attention to the relationship between trade and the environment (as well as trading with the environment) is gaining increasing attention (see for example Ratnayake 1999). New Zealand also relies on its indigenous natural resources to fuel its economic growth in other ways. The mountainous elevations, glacial past and sloped terrain combined with abundant rainfall offering potential for hydroelectric power and the majority of the nation's electricity (and New Zealand has one of the highest per capita consumption of energy in the world) comes from this source. But while rural industries such as farming and horticulture are still important it is tourism that has become the primary source of export earnings. In 2004 international tourism generated 18.5 per cent of export earnings compared to export receipts from dairy products of 14.3 per cent (Te Tari Tatau 2005). New Zealand's tourism product is diverse, but it is predominantly reliant on outdoor and scenic activities, which has led to the 'clean, green' moniker being used in tourism and policy literature (Bührs 1993; MfE 2001; Stone 2003). Therefore, as in the past, natural resources remain the cornerstone of New Zealand's society and economy.

Nature, Culture and Society

Although the prehistory of New Zealand is not definitively understood it is generally believed that the islands remained largely unoccupied until around 800 years ago (McKnight 1995). The first true settlers were migratory Polynesians and waves of immigration occurred over time such that what is now known as the Māori culture was well established on the North Island by the time the first European settlers arrived on the islands late in the 18th century. The first European settlements were sealing and whaling stations on the west coast of the South Island and it was not until 1840 that the first colonization settlement was established at Wellington by a British enterprise, the New Zealand Company. Within a week British sovereignty was declared over New Zealand and Captain William Hobson of the British Navy (later the first Governor of New Zealand) and a group of Māori chiefs signed the Treaty of Waitangi that declared British rule over the North Island (the South Island was claimed by the British as a right of discovery). As mentioned in the previous section the aim of the treaty was to offer Māoris protection of their rights, including property rights, in return for their acceptance of Queen Victoria as their ruler. There were, however, two versions of the treaty, one in English and one in Māori, and disputes regarding the translations of key words such as stewardship, sovereignty and governorship have been on-going debates since its declaration.

After the signing of the Treaty of Waitangi the European population grew and settlements expanded. Initially this occurred more rapidly in the South Island because of lower levels of Māori population, and hence less conflict with the indigenous population, but the expansion was also fuelled across both islands by the development of refrigerated shipping in the late 1880s that facilitated the exportation of meat and dairy products. This technological advancement also accelerated the clearing of large areas of dense forest and bush for farming. Before human settlement it has been estimated that three-quarters of the islands had forest cover and while 50 per cent was still forested by the time the European settlers (Pākehā) arrived this was reduced to a quarter by the early part of the 20th century (Sturman and Spronken-Smith 2001). Human hands then heavily influenced the landscapes now described by the New Zealand government as paradise and as Wynn (2002) reports the process of deforestation has been perceived as both destruction and improvement of nature. There are those who saw landscape changes as the natural order of things where nature '... had been obliged to yield to intelligent human guidance' (Sargenson 1981, 53). In this reading deforestation was seen as a process of civilization for wild nature and a great achievement of the European settlers. As Dunlap (1999) reports, the settlers dream was predominantly one of remaking the land, 'the settlers destroyed and re-created, appreciated the beauties of the land, and sought to bring it closer to their own ideal, and they did it on a grand scale' (1999, 46). Yet there were alternative perspectives including Scholefield (1909) who in the early 20th century wrote with sadness about the war that had been waged against the forest in the name of progress and pastoral perfection.

New Zealand has then been shaped by activities directly related to European imperialism that sought to colonize new territories and bring them into the capitalist world economy. As a result relations between humans and nature (or environments) in New Zealand have parallels with other locations that experienced colonization, but the context for those interactions and the rapid timescale over which they occurred remains unique (Pawson and Brooking 2002). In particular the role of Māori in both effecting environmental change and challenging the environmental changes brought about by European settlers is beginning to achieve significant attention as an important part of environmental history. Both Anderson (2002) and Stokes (2002), for example, seek to dispel the commonly held myths of Māori as either environmentally benign noble savages or incapable of environmental change, highlighting the misunderstandings that took place between Māori or Pākehā over notions of land ownership, territoriality and boundedness. These misunderstandings led to conflicts over landscapes affected by activities such as mining where excavation led to conflicts over land use, siltation of rivers and flooding (Hearn 2002), but they also facilitated the early establishment of environmental legislation such as the New Zealand Forests Act of 1874. Although utilitarian valuing of indigenous resources drove much of the early environmental legislation, by the beginning of the 20th century there was a more defined notion of New Zealand as a home-place (Star and Lochhead 2002).

Into the 20th century environmental transformations were directed by scientific and technical developments that facilitated the intensification of land use. This intensification caused those concerned with what became known as wise-use of resources to see farming as a war against nature. It was also a time of urbanization for many New Zealanders, including Māori, and the emergence of the suburb, and more precisely gardens and gardening in the suburbs, has become an important site of nature-society interactions (Leach 2002). But perhaps most significantly for this analysis it was also a century of legislation for both resource use and, more recently, resource conservation. The early pattern of legislative development was for the Crown to assume ownership of resources and then reallocate rights to private users (Wheen 2002). For the most part resource conservation matters were only introduced where health, safety or economic interests were at risk. Indeed Wheen describes legislation for conservation as 'mere ripples in the blanket of law enabling resource development' (2002, 262). Nowhere was this more visible than in the development of hydroelectricity in the post World War II era. As pressures mounted to industrialize a proposal was set out for the development of a hydroelectric facility at Lake Manapouri in the South Island. However when the private developer failed to progress the initiative the government pushed through an act that reverted all water rights to the Crown enabling the state to develop the resource. It was only after the construction of the facility in 1969, and following the first mass mobilization of public concern over an environmental issue, that the impacts of the development were considered. The original act was amended in 1981 and guidelines for the protection of ecological stability and recreational values were included as relationships between conservationists and the industry were established. The Manapouri case elevated the politicization of environmental issues in New Zealand and has been seen as stimulating the introduction of cost-benefit analysis and environmental impact assessments. Wheen (2002) and others have argued that a bias towards resource development rather than environmental protection has persisted particularly during the difficult economic times that occurred following the loss of the UK market for agricultural products during the 1970s and the recession of the 1980s. Yet New Zealand is renown for having one of the strongest environmental protection regimes built around the 1991 Resource Management Act. The evolution, importance and impact of this legislation is detailed in the following section but it is important to emphasize here how it has played a pivotal role in perpetuating the clean and green image that has common currency amongst tourists to New Zealand and for New Zealanders themselves. Although it has been established that this notion of paradise, characterized through concepts of cleanliness and greenness, draws on a particular historical colonial place myth, it has been reinforced through more recent policy positions in relation to nuclear power and environmental protection (Bell 1996; Coyle and Fairweather 2005; Dew 1999; Shields 1991). In addition the clean green branding has been used as a driver for debating legislation to promote more sustainable practices in the face of pressures threatening environmental degradation (Fleming 2002).

The existence of pro-environmental attitudes is fairly well distributed across socio-economic groups, ages and ethnicities in New Zealand with significant concerns expressed about environmental issues in general surveys, with waste being seen as one of the most challenging areas (Hughey et al. 2004; Massey University 2001). Nonetheless public opinion surveys also reveal that New Zealanders generally still consider the state of their environment to be good and better than in other developed countries. This is despite the increasing evidence from environmental science and monitoring that 'clean and green' is more myth than reality in contemporary New Zealand (Taylor 2005). In contrast to the words of the Environment Minister that opened this chapter Christine Dann responds

> the creeks run high and brown, and flash stream course straight down steep, bare hillsides, taking with them the yellow loess soils that the roots of totāra used to bind to the slope. On the hill opposite the old cook's quarters, the dead and dying kānuka trees thrash in the wind. Helicopter spraying killed them along with some gorse. The kāwhai on our side of the creek also copped the spray. They are still alive but won't flower this spring. The creek rampages by, carrying dead trees and dead sheep. The head of the harbour is already smothered in fine silt as deep as my thighs, and I see the spreading brown stain from the hills stretch out to the open sea, carrying still more mud to settle. The depleted cockle beds in the harbour are currently protected from plundering by a special five-year rāhui, but what shellfish can withstand this slow death by suffocation as hills keep sliding into the sea? (2002, 275).

In essence New Zealand's story is one of the environmental impacts of human settlement and development, particularly the settlements and developments that were established by the Pākehā through intertwined processes colonialism, industrialization and modernization,[4] but how have these processes affected the environmental governance and particularly the treatment of waste? As a first step to answer this question the next section considers the emergence of environmental policy in New Zealand.

Environmental Policy

As indicated above, the end of the twentieth century saw a period of restructuring in New Zealand that permeated social, economic and environmental governance arenas. These changes had a strong geographical dimension to them both through the scalar restructuring of responsibilities for planning and resource management and through the redefinition of human-environment relations (see Britton et al. 1992; Cocklin and Furuseth 1994).

From an international perspective New Zealand has developed a reputation as an innovator in the environmental field being the site of the first national green party and by creating pioneering environmental institutions such as the establishment of an independent agency to evaluate the effectiveness of environmental legislation, the Parliamentary Commissioner for the Environment (Bührs and Bartlett 1993;

---

4   Dann argues that far from experiencing a period of post-colonialism New Zealand retains an entirely colonial approach to environments through the introduction of alien agricultures and exotic biota and a commitment to a rights based approach to ownership of resources. Industrialization in New Zealand is related to the intense application of industrial methods to farming, primarily, and practices that effectively treat the land as a factory and often have insidious and invisible impacts on environmental quality.

Garner 2000). These innovations were stimulated in part by the strength of environmental activist campaigns against large scale developments such as damn construction, scrub clearance and the logging of indigenous forests that occurred from the 1960s. These developments generated an adversarial relationship between environmentalists, developers and the state (who were often the sponsors of the developments) to the extent that environmental issues became highly politicized (Wilson 1982). The response during the 1980s, in line with other neo-liberal reforms of government practice, was to retreat from direct involvement in resource management transferring the challenge of dealing with contested developments to the private sector – for example state owned forests not designated as being of high conservation value were privatized – or to local government and the courts (Bührs 2003). Issues relating the property rights, significantly both of Māori and Pākehā, and the integration of economic principles into environmental management typified the drivers behind these periods of change and both reflect the neo-liberal, free-market ideology visible in other sectoral restructuring processes. The context for restructuring of environmental management then was affected by internal (domestic) and external (international) pressures that had economic, social and environmental dimensions as illustrated in Table 4.1.

Table 4.1 Pressures for environmental restructuring in New Zealand

	Subnational	**National**	**International**
Economic	• Regional economic disparities	• Domestic economic crisis	• Internationalism • Crisis of accumulation
Social	• Regional decline • Issues of representation	• Widening disparities • Cultural identity	• Social effects of environmental change e.g. health
Environment	• Regional and local effects of industrial development • Resource exploitation	• Pressure for administrative reform • Resource development concerns	• Sustainable development • Global environmental change

Source: Adapted from Cockerill and Furuseth (1992).

In essence the reform of environmental administration during the 1980s can be disaggregated into three phases, the first saw the reorganization of central government environmental agencies, the second addressed the form and functions of local and regional government and the third developed new legislative arrangements, specifically the Resource Management Act (Cocklin and Furuseth 1994). While the Local Government reforms of the 1980s restructured the administrative boundaries of local authorities and addressed the remits of the newly formed administrative units, further changes aimed at moving towards local sustainability were drafted into the 2002 Local Government Act. In the Act each local authority is required to produce Long Term Council Community Plans (LTCCP) by 2006 setting out the goals for the community over the next decade, to be revised every three years. Financial costing and targets were identified as key elements of measuring progress towards goal implementation in these plans. The impact of these plans is hard to ascertain given their relatively recent adoption and to date the 1991 Resource Management Act (RMA) remains the defining feature of environmental planning in New Zealand.[5]

The RMA was formed after considerable consultation between stakeholders and replaced over fifty existing Acts covering the use, development and protection of air, land and water (Wheen 2002). It established a multi-scalar legislative system incorporating national, regional and local scales of government in an attempt to overhaul a system criticized for its complexity and inadequacy. The Act demands that the actual and potential effects on the environment of any proposed development must be assessed and a wide range of interests considered (including those covered by the Treaty of Waitangi) before a development or activity can proceed. Underpinning these assessments is a stated commitment to promote sustainable management and it was this early acknowledgement of sustainability issues that led to New Zealand being heralded a leader in environmental legislation. Sustainable management was defined by Section 5 of the Act as

> the use, development, and protection of natural and physical resources in a way, or at a rate, which enables people and communities to provide for their social, economic and cultural wellbeing and for their health and safety, while – a) sustaining the potential of natural and physical resources (excluding minerals) to meet the reasonably foreseeable needs of future generations; and b) safeguarding the life-supporting capacity of air, water, soil and ecosystems; and c) avoiding, remedying, or mitigating any adverse effects of activities on the environment.

However, rather than developing strict standards or rules defining what this sustainable management might mean in practice, appeals brought under the Act produced a broad reading of sustainable development that simply called for a balanced consideration of environmental and economic costs and benefits.

From its inception the RMA attracted considerable attention from policy makers and academics across the globe. Initially heralded by some commentators as a pioneering form of integrated statutory sustainability planning and management (see Memon and Gleeson 1995) there has been some reflection on the limitations of the RMA. Three main areas of concern have been raised relating to a lack of central government guidance on implementation, a marginalization of social concerns and a de-politicization of environmental decision making in what remains a highly contested arena (Bührs 2003; Grundy and Gleeson 1996; Murray and Swaffield 1994; Perkins and Thorns 2001; Walker et al. 2000). What is unquestionable however is that the RMA fundamentally changed the foundations of environmental management in New Zealand. Consolidating disparate environmental planning statutes the RMA

5　Detailed procedural information about the RMA can be found in Memon and Perkins (2000), Peart (2004), Erickson et al. (2003).

established a system of plans and policy statements at the territorial and regional level. Within these plans certain activities are permitted, but those which are not are required to apply for resource consents from the relevant territorial (for land use issues) or regional (for discharges to air, water and land) governments. Applications are required in the form of environmental statements documenting the predicted effects of the activities on the environment. In this way the RMA moved planning from a zone-based system, similar to the British town and country planning system, to an effects-based framework that was predicated on permitting any development provided it did not have adverse impacts on the biophysical environment and upheld the sustainable management of land, air and water (Perkins and Thorns 2001). A presumption in favour of development, as long as negative effects can be dealt with satisfactorily, is enshrined within the RMA. As stated in the Ministry for Environment's guide to the RMA 'the underlying assumption is that any use, development or subdivision should proceed if there are no adverse environmental effects, or if these effects can be avoided, remedied or mitigated' (MfE 1999, para 1.4). During the formation of the RMA the Minister for the Environment clearly stated that the overarching aim was to provide a liberal regime for developers on the understanding that a physical bottom line would not be compromised (Upton 1991).

The RMA devolved much of the discussions and decisions about the effects of developments to the local level and to the courts. It is the regional and territorial authorities that, according to their statutory remits, in the first instance decide whether a development should be permitted or not based on an assessment of its predicted environmental effects in relation to local plans. District and regional plans are an important aspect of the RMA. Councils use plans to set out how they will protect the local environment and they are publicly available at councils and libraries. Regional plans, prepared by regional councils (for their region), mostly relate to the coast, rivers, soil and the air. They set out how discharges or activities using these resources will be managed to stop the resources being degraded or polluted. District plans, prepared by district and city councils (for their district or city) focus on issues relating to the use of land. This includes looking at locally valued species, habitats and environments such as trees, forests, farm land and even suburban areas. The plans set out how land use and development activities will be managed to protect those values.

Where activities may affect the environment significantly applications for resource consent are usually necessary. Things that may require consent include putting waste into a stream, taking groundwater, subdividing land, or building a garage, but these vary from place to place as different councils are able to set their out their own requirements. Once a proposal for resource consent has been submitted parties can make submissions in relation to it. These submissions can be supportive, neutral or negative and can be made by any party, not just the local community or directly affected parties.

There are channels for appeal against the decision of the local authorities that lead appellants to the Environment Court and commentators have praised the system for its participatory mechanisms and its co-operative structures (Berke et al. 1999). However both the local authority decision making process and the appeals procedures are tightly bound by legal and scientific definitions of sustainability and rely heavily on the input of expert witnesses (Ong 2001). Indeed as Bührs states

> whether development proposals are considered to be desirable or acceptable from a broader environmental, ethical, social, economic or political perspective is not considered to be a legal (or even legitimate) ground for decision making (2003, 94).

So while the RMA has fairly extensive provisions for public participation it has been argued that the procedures tend to favour those who have greater resources to employ expert witnesses and can afford to take the financial risk of challenging decisions (Chapple 1995; Gunder and Mouat 2002). The onus is very much on individuals to act according to legal requirements within strict timetables. Failure to follow these requirements can exclude participation at later stages of the planning process.

Despite attention to the Treaty of Waitangi in the RMA the tendency for Māori communities to fall within lower socio-economic classifications means that such a legalistic and science-based system has been linked to the potential exclusion of indigenous populations from participating in the protection of environmental integrity (Berke et al. 2002; Lane 2006). In addition there have been concerns expressed about the number of applications that are subjected to public scrutiny with local authorities using extensive powers for non-notification and with developers apparently able to buy off objections in order to avoid conflict (Gunder and Mouart 2000). In the name of administrative efficiency the RMA allows non-contentious resource consent applications to be approved without public notification because they comply with permitted activities as detailed in plans, are considered to pose only minimal effects or have been agreed to (in writing) by those parties who might be adversely affected by the development. In 2000, Gunder and Mouat reported that around 95 per cent of all resource consents were not being publicly notified. At the same time developers can, at their discretion, offer compensation to affected parties in order to 'foster good will between developers and affected party; to reduce the uncertainty of the consent process; and to avoid or reduce costs' (Bevan and Jay 1998, 4). Provision for such compensation remains despite recognition from the Ministry for the Environment that the process is open to abuse and has the potential to permit unethical practices (MfE 1996).

As detailed previously the specific form of the RMA emerged out of a context of neo-liberal reforms of the Labour government during the 1980s as well as out of long negotiations between environmentalists and business (Kelsey 1997; Le Heron and Pawson 1996; Memon and Gleeson 1995). The solution was to create a system based on scientific knowledge of ecosystems and an economically informed understanding of costs and benefits. In theory the RMA provided the architecture through which the environmental effects of proposals could be compared against precise and pre-set environmental standards. Market forces would then attend to the allocation of resources with respect to permitted developments. However such a process requires the existence of precise, objective and universal natural environmental standards unfettered by the vagaries of political, social and ethical judgements. According to Perkins and Thorns (2001) the focus on biophysical interpretations of the environment and the adoption of market-based vocabulary under the guise of sustainable management creates a veneer of objectivity over what remains an essentially political process of determining the effects of developments. In addition it has been suggested that it may be unreasonable to expect local planners, in terms of both skills and resources, to identify all the environmental and socio-economic effects of planning options and produce a plan that provides an adequate baseline against which to measure applications for resource consents. The decentralized system of analysing effects for specific development activities also means that cumulative and synergistic impacts are not easily addressed in consenting procedures, even though this is required by the RMA (Day et al. 2003).

Decentralization has been linked to problems of inconsistency in the application of the RMA in the absence of strong national policies or guidance and the lack of a dedicated pollution control agency. Although there are demonstrable resource constraints for the Ministry for the Environment and allied agencies, such as the Parliamentary Commissioner for the Environment, Bührs (2003) suggests the weakness of the environmental cause within national government is centrally related to the mandate of the Ministry for the Environment. Despite its title the Ministry is obliged to have regard for the full range of perspectives when advising on environmental policy rather than promoting environmental interests or values. As a consequence it 'rarely takes a strong stance on environmental issues, or 'rocks the boat' in terms of advocating radical policy changes that could threaten dominant interests' (Bührs 2003, 95).

The evolution of the RMA and its subsequent implementation has occurred alongside local government reform. As mentioned previously local government is a creature of statute such that local government structures, competencies and processes are constructed through local government acts created at central government level. In 1989 the Local Government Act 1974, a seminal act for local government practices in New Zealand, was amended significantly to reduce the number of local authorities, to require mandatory strategic plans and to introduce more business-based approaches (Perkins and Thorns 2001). The business-led approaches were further consolidated following the 1996 Amendment that introduced a strict economic allocation model for the construction of strategic plans within local authorities (Welch 2002).

From a governance perspective it has been argued that the RMA reflects wider neo-liberal movements by the state to stand back from central government policy making. As the central state withdrew the challenge of balancing economic development and environmental protection fell primarily on the shoulders of local and regional governments who were provided with little in the way of formal guidance about how to proceed (Cockeril and Furuseth 1994). While this was seen as a means to allow local flexibility in providing solutions to environmental challenges it was always possible that the variations between regions in terms of environmental, economic and social resources could lead to unequal practices. At the same time the focus on scientific and technical assessments, the language of efficiency allied with privatization of resources and devolution of decision-making responsibilities was seen as an attempt to de-politicize and de-ideologize environmental decision making, while allowing dominant economic interests to maintain their influence. Bührs (2003) claims that since the late 1980s with the introduction of the RMA the relationship between the different views across the environment-development spectrum 'appears to have become less conflict-ridden, and more accommodating and even co-operative' (2003, 98). Yet, as noted by Perkins and Thorns (2001), the consent process remains a site of conflict between environmentalists fearful of overly liberal allocation of consents and developers who are frustrated by what they see as unnecessary regulation. So while New Zealand was a forerunner in articulating the principles of sustainable development in its planning system there remain concerns about whether this rhetoric has been matched by effective implementation (Berke et al. 2006).

The drafting and implementation of the RMA was followed in 1995 by the publication of New Zealand's first strategy for sustainable development 'Environment 2010' with the vision of creating 'a clean, healthy and unique environment, sustaining nature and people's needs and aspirations' (MfE 1995). The principles of the strategy focused on integrating environmental, societal and economic considerations and waste management was identified as one of the identified priority areas for action. One of the main comments in the strategy was the lack of consistent national environmental data or standards to indicate the quality of the environment and the waste sector was no exception. The MfE found that data on waste was limited in coverage and accuracy and concluded that waste was increasing, polluter pays principles were not being applied successfully to waste producers and that landfill standards were underdeveloped (MfE 1998). This was perhaps to be expected given that, with its focus on effects, there is little specifically that can be done through the RMA about the creation, reduction or prevention of waste. The RMA is only concerned with reducing or minimising the negative effects on the environment from waste produced by developments. For example landfills, incinerators or waste transfer stations, may need to be considered through the RMA process but only because of the impact their activities may have on the environment through emissions to air, land or water. Whether such facilities should exist in the first place is not considered a relevant issue (Brodnax and Milne 2002).

The Parliamentary Commissioner on the Environment raised concerns about the lack of progress on sustainable development in a review conducted in 2002 (PCE 2002) and this stimulated the production of a programme for action on sustainable development (Department of Prime Minister and Cabinet 2003). The programme charted the challenges and opportunities for New Zealand and linked sustainable development to innovation and collaboration. Central to this national government commitment to sustainable development was recognition of the need for leadership in articulating outcomes and directions for New Zealand, but little was said in this document about waste management.

As with Ireland the environmental policy landscape in New Zealand raises considerable questions about the role of science and public participation in environmental issues. However it is clear that there are differing environmental governance styles between the two countries based on the particular development patterns and nature-society relations that are unique to each nation. The issue then becomes how to conduct a comparative governance analysis of these two cases. A reflection on some methodological considerations of such comparative governance analysis is presented below.

Methodological considerations

In the past governance studies have been criticized for their lack of empirical foundation (Jordan 2001), yet even when there is a solid empirical basis for governance studies there has been little explicit reflection, with the notable exception of Rhodes (1997), on the methodological basis for that work. Some methodological components of an empirically grounded governance analysis may be unproblematic, in the sense that formal governing structures, or the hard infrastructure of government (Healey 1997), are usually enshrined in publicly accessible policy documents that are readily available for textual or discourse analysis. It is a more challenging endeavour to identify and examine informal structures and interactions between actors and agencies, particularly across scales, cultures and spheres of governance. This work does not provide a conclusive solution to these issues, rather it seeks to be transparent about the basis for the statements made in the text.

Following the ideas of Yin (1984) and Eckstein (1975) the aim was to bring description and analysis together productively in order to facilitate analytical rather than statistical generalizations. According to Rhodes (1997) the comparative case method allows valid generalizations when a theoretical framework is used to structure the case studies and is particularly useful when investigating complex phenomena. In sum the method sought answers to both the 'what-questions' and the 'why-questions' of governance; to combine both description and analysis through a comparative case study method.

Of course there are some challenges when conducting comparative research of any kind. Researchers frequently report problems with managing and funding cross-national projects as well as barriers resulting from linguistic or cultural differences. Most significantly in this research however were difficulties in gaining access to comparable datasets and establishing functional equivalence between actors and agencies to capture the complexity of the cases whilst also wishing to produce some level of generalization in terms of comparison. These challenges were addressed through careful investigation of the national contexts and through negotiation and compromise in data collection practices. For example, it was not possible to examine and report on every relevant document, interview representatives from every relevant institution and perspective, or cover every development in the dynamic and complex waste policy processes in both countries. Instead a process of survey and selection took place in order to identify significant moments, actors and positions. In each country the collation of secondary evidence in the shape of previous research and published documentation was considered and this was used as the starting point from which to identify individuals for in-depth, qualitative interviews. In addition to comprehensive policy analysis more than sixty interviews were conducted across the two case studies and actors from across the governance spectrum were involved in the process of data collection.

Based on the textual analysis and the interview data a tripartite framework for comparative governance analysis was developed and the findings are reported in the following two chapters. The first stage in the process was to identify and examine the evolution of waste policy interventions that includes waste policy initiatives, regulations and programmes. These interventions effectively make up the waste policy landscape for both countries. However examining the landscape without an appreciation of the processes that occurred to form that landscape would tell only a limited story of the governance of municipal waste. In recognition of this the second phase involved an examination of the processes, practices and negotiations between different actors, agencies and organizations, operating at and across different scales, which led to the policy interventions. The third phase reflects on the outcomes of the interactions and subsequent interventions embedded within dynamic social, environmental, political and economic contexts.

Chapter 5

Garbage Governance in Ireland: Waste Wars in the Emerald Isle

Introduction

The governance of waste is widely held to be one of the most problematic areas of environmental management in Ireland with high profile conflicts over many issues from landfill expansion to charges for municipal waste collection. These tensions have been created as a result of rapid increases in waste generation across all sectors since the upswing in the economy of the late 1990s and through increasing demands from EU Directives to divert waste away from landfill, the dominant mode of waste management in the Republic. This chapter provides an examination of municipal waste management that is set within the wider socio-political context of environmental governance in Ireland as outlined in the previous chapter. The waste governance analysis is structured to attend to policy interventions, interactions between governance actors and governing outcomes. Following Bulkeley et al. (2005) consideration is given to initiatives that have created the waste policy landscape in Ireland, the rationalities that underpin those initiatives, the governing agencies involved and the mechanisms (or tools and technologies) through which those policy interventions have found articulation. This information provides the backdrop for an examination of the interactions between actors and structures that have formed and reform the resultant mechanisms and policies. Combining the results of the preceding analyses the third section reflects on the impacts that the particular geographies and interactions have had on waste governance in Ireland. Finally a case study of the struggles between different interests, individuals and organizations over the adoption of waste management plans is used to illustrate the complexity of governing processes and the multiple sites of governing that occur within the waste field.

Policy Interventions

As a core function of local authorities with no direct involvement from central government, an underdeveloped regulatory framework and no external regulation of local authority waste services, waste management was fairly representative of Ireland's environmental policy during the 1970s and 1980s. The waste system comprised of basic, unseparated collection services and landfills (or town dumps as they were more commonly known) for municipal waste and virtually no recycling of waste. Indeed Ireland had one of the lowest recycling rates in the EU, no facilities for biological treatment and no capacity for municipal waste to energy recovery through incineration. From a policy perspective waste was predominantly managed through a number of public health statutes and ministerial regulations based on the demands of relatively loose European legislation (Meehan 1996). However by the mid 1980s waste was becoming a higher priority within the EU and Ireland was encountering more external pressure to adhere to stricter Directives. At the same time the government was facing increasingly voluble public protest against poorly managed local dumps to the extent that 'the problem of waste management emerged as one of the most politically contentious areas of environmental politics in Ireland' (Taylor 2001, 97).

From an institutional perspective a pivotal moment in Irish waste management was the establishment of the *1996 Waste Management Act* that incorporated demands for management plans, licensing procedures and increased monitoring of implementation and compliance (see Table 5.1). Effectively it marked a first attempt to develop a comprehensive national framework for waste management strategies. Within the Act the production of waste plans by local authorities was seen as the key mechanism by which the strategic management of waste could be developed. Following the demands of EU legislation[1] the main aim of these plans was to reduce the amount of waste going to landfill, which at the time accounted for around 98 per cent of domestic and 70 per cent of commercial waste (Taylor 2001). The Act ushered in a new regime for waste management activities with more clearly defined regulatory roles for local authorities and the EPA, a qualified role for local authorities in terms of service provision and an acknowledgement of the private sector's potential contribution to waste management.

The legislative developments of the 1996 Act were further extended with the publication of the Government policy statement on waste *Waste Management: Changing Our Ways* (DoELG 1998) which was primarily concerned with the reduction of disposal to landfill with increased recycling and recovery of waste. Targets for waste management were set for the first time with an objective of diverting 50 per cent of household waste and 65 per cent of biodegradable waste from landfill by 2013. In addition 35 per cent of municipal waste and 85 per cent of construction and demolition waste was to be recycled. The participation of the private sector, either directly or through public private partnerships, in achieving these targets was emphasized and authorities were encouraged to facilitate business involvement in waste management service on the grounds that

> [p]rivate participation can contribute much needed capital investment in infrastructure, specialist expertise in the application of alternative and emerging technologies, a better understanding of the dynamics of the marketplace ... [and] it can also release local authority staff and resources for other productive uses (DoELG 1998, 8).

---

1   Under the EU's interpretation of sustainable waste management, and through the attendant waste management hierarchy, disposal to landfill is the least favourable option, followed by thermal waste treatment (with waste to energy transfer), recycling, re-use and waste minimization, with prevention the most favourable position (EPA, 2000: 1). The EU Landfill Directive set targets for the diversion of biodegradable waste from landfill 75 per cent of 1995 levels by 2010, 50 per cent of 1995 levels by 2013 and 35 per cent of 1995 levels by 2020.

Table 5.1 1996 Irish Waste Management Act

Part	Component Description
I **General provisions**	New powers for local authorities to facilitate enforcement of the legislation through inspections and through obligations on waste collectors for monitoring and information provision with attendant fines for non-compliance
II **Waste management planning**	EPA is to provide a hazardous waste management plan while local authorities are to provide detailed waste management plans that are to reflect to EU waste management hierarchy and are to be reviewed every five years. A two-stage consultation is required, one period before the preparation of the plan commences and one at the draft stage
III **Waste prevention and recovery**	To include an obligation on industrial, commercial and agricultural activities to have due regard to waste prevention and recovery
IV **Holding, collection and movement of waste**	Permits required by commercial waste activities. Empower local authorities to develop bye-laws for presentation of waste
V **Waste recovery and disposal**	Local authorities have a duty to ensure that there are adequate facilities for the recovery and disposal of domestic waste
VI **General provisions regarding environmental protection**	Provisions to introduce measures to prevent, limit or remedy the effects of environmental pollution caused by the holding, recovery or disposal of waste

Source: Adapted from the Irish Statute Book 2006.

Local authorities proceeded to develop their waste management plans with all bar three opting to form regional plans in collaboration with neighbouring areas (see Figure 5.1 and Table 5.2). Only Kildare, Wicklow and Donegal drafted plans opted to draft their own county-based plans and in each of these cases there was the view that future regionalization might occur through a cross-border plan between Donegal and Ulster and through Kildare and Wicklow joining with the Dublin region. Such regionalization had been encouraged by central government as a way of developing a more efficient provision of services and infrastructure. Efficiency here was directed towards economies of scale that, it was assumed, would provide a viable framework in planning and volume terms for the development of integrated and innovative waste management solutions as well as fostering a positive climate for public private partnerships. A degree of local autonomy was retained in this regional structure however as all local authorities within a region had to formally adopt a plan through a full council meeting of elected officials for the process to be complete.

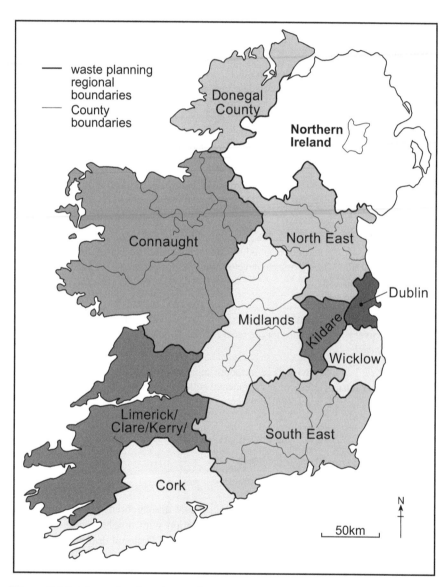

Figure 5.1 Irish waste planning regions

Table 5.2 Irish waste planning regions

Region	Councils
North East	Cavan, Meath and Monaghan
Dublin	Dun Laoghaire-Rathdown, Fingal, South Dublin, and Dublin City Council
Midlands	Laois, Longford, Offaly, North Tipperary and Westmeath
Connaught	Galway, Leitrim, Mayo, Roscommon, Sligo and Galway City Council
Limerick/Clare/Kerry	Clare, Kerry, Limerick and Limerick City Council
Cork	Cork and County Cork
South East	Carlow, Kilkenny, South Tipperary, Waterford, Wexford and Waterford City Council

Notes: Donegal, Wicklow and Kildare did not join a regional structure, but made individual county waste management plans.
Source: Adapted from Davies (2003, 81).

Central government advised local authorities that they should seek assistance from experts in the field of waste management to draw up their waste plans and engineering consultants were duly brought in. Fehily, Timoney & Co. produced the Joint Waste Management Plan for the South East, Tobin Environmental Services Ltd. were the consultants for the Cork and M.C. O'Sullivan and Co. Ltd. (MCOS)[2] prepared the remainder of the plans. These consultants acted as educators and information providers to local authorities, industry, business and publics during the waste planning process and as such they were highly influential in waste management debates and pivotal in defining the strategic vision for future waste planning across Ireland. Due to the directions from the Minister for the Environment to consider integrated waste management, and the dominance of a few key consultants, the plans produced were remarkably similar both in presentation and content, with each regionalized plan recommending thermal treatment alongside recycling, biological treatment and reduced landfill.[3] Specifically it was their identification of the need for municipal incineration facilities that was to cause the most problems in the process of adopting waste management plans. Indeed so challenging did some local authorities find reaching agreement about incineration that in 2000 six had not adopted the plans drawn up by consultants. During 2001 Europe was beginning

---

2   MCOS were taken over by the international environmental consultancy group RPS in 2002.

3   While the single authority waste management plans did not propose to build incinerators within their boundaries due to economic constraints they did not rule out the possibility of reconsidering incineration at a later date.

to lose patience with the lack of progress in adopting waste management plans in Ireland and stated that if no action was forthcoming then the state would be taken to the European Court of Justice for non-compliance with the 1999 EU Landfill Directive. This threat of intervention from Europe created tensions between central government and those local authorities that had not adopted plans. Central Government was keen to move the waste planning process forward in order to avoid confrontation with Europe, but it was also unwilling to renegotiate the content of the plans in the face of resistance from locally elected councillors who had the responsibility for adopting the plans.

Eventually the Minister for the Environment intervened in the stalemate when he introduced a Bill to amend the 1996 Waste Management Act. Alongside an environmental levy both on plastic bags and waste sent to landfill the Act transferred the responsibility for the adoption of waste management plans from the elected members of local authorities to local authority managers. As noted by Boyle (2001) a clear aim of this transference was to remove the adoption decision from the electoral process, but by doing so central government lay itself open to criticism for eroding fundamental aspects of local democracy. It appeared to some, particularly opposition politicians, that the Minister for the Environment simply wanted plans adopted irrespective of the local appropriateness of the strategies contained within them. In this regard the Minister achieved his aim and all waste plans were adopted in their original state by the end of 2001.

Following the 2001 Amendment a newly appointed Minister for the Environment, Martin Cullen, moved the debate regarding the process of waste management planning to another stage. As mentioned previously, in 2002 he launched a proposal to circumvent local planning processes by fast-tracking decisions about large scale infrastructure projects, including waste infrastructure such as incinerators and landfills, directly to An Bord Pleanála (the Irish Planning Appeals Board). Coverage of this proposal focused on his labelling of the waste management planning process as 'over-democratized' and opposition politicians were quick to label the move anti-democratic fearing a sense of disempowerment for local communities and a further reduction in their input into waste management processes. In the face of considerable resistance it was not until February 2006 that the Government published a formal bill on the issue, which was finally brought into force as the Planning and Development (Strategic Infrastructure) Act 2006 in July of that year in order to

> provide, in the interests of the common good, for the making directly to An Bord Pleanála of applications for planning permission in respect of certain proposed developments of strategic importance to the State; to make provision for the expeditious determination of such applications, applications for certain other types of consent or approval and applications for planning permissions generally; for those purposes and for the purpose of effecting certain other changes to the law of planning and development to amend and extend the Planning and Development Acts 2000 to 2004; to amend the Transport (Railway Infrastructure) Act 2001 and the Acquisition of Land (Assessment of Compensation) Act 1919 and to provide for related matters (Houses of the Oireachtas, Bill Number 27 of 2006, 5).

Interestingly, at the launch of the original Bill the Minister for the Environment acknowledged that most of the delays to decisions on infrastructural developments have been due to legal challenges to planning decisions (made by An Bord Pleanála) rather than due to local authority processes (Forfás 2006, 28). Critics suggest that without attention to these legal processes it is unclear what impact the Act will have on speeding-up planning processes.

At the same time as discussions about fast-tracking waste infrastructure developments were initiated a policy statement focusing on waste prevention and recycling was produced – *Preventing and Recycling Waste: Delivering Change* (DoELG 2002). Its aim was to move action in the waste management field further up the waste management hierarchy by modernising the recycling infrastructure and proposing greater producer responsibility initiatives. This was followed by the Protection of the Environment Act (2003), which outlined stronger enforcement provisions in the waste field and updated planning, licensing and permitting procedures. Specifically the Act stated that in addition to the adoption process the review, variation and replacement of the waste management plans should also be made by the local authority manager and that in the case of a conflict between the objectives of a development plan and those of the waste management plan, the waste plan should take precedence.

Key to the development of initiatives to support waste management is the provision of funds. The Irish government provided resources from the plastic bag and landfill levies that were both developed following the amendment to the 1996 Waste Management Act. The money raised was ring-fenced and it became known as the Environment Fund. The money was used to construct further recycling facilities through bring banks and civic amenity sites as well as providing support for litter prevention programmes and waste awareness initiatives. Initially the grant allocations were deliberately targeted at recycling initiatives that would be visible at the local level in order to help raise awareness and to demonstrate the government's commitment to supporting recycling development and since November 2002, more than €90 million has been provided to assist local authorities with the capital costs of providing a range of local authority projects. One key feature funded was the *Race Against Waste* campaign, supported by the Department of Environment, Heritage and Local Government and managed by RPS- MCOS, in 2003. It was the first time that a national environmental awareness campaign had focused on a single issue and it indicated the high profile the sector had attained. The multimedia information campaign included the development of fact sheets in Irish and English, a web-site, television and cinema advertising and a low cost information hotline for publics, small businesses and the public sector. Evaluations have claimed that the campaign was successful in 'supporting a new attitude to the waste we produce and identified a willingness of individuals to participate in waste reduction and recycling measures' (DoEHLG 2004, 1), but it was recognized that much more change would be necessary over the long term. There were also concerns raised about the high cost of the initiative (€3.4 million over two years) given the difficulty of measuring any direct impact from a national campaign. The initial campaign was followed by a similar cross-boarder initiative part funded by Interreg in 2004. However the evaluation of this project illustrates the difficulties of using market based techniques to stimulate behavioural change. The success of the project was primarily measured in terms of how many people remembered seeing the advertising. A subsequent post-campaign survey did report that 23 per cent of respondents reported that they had made an attempt to reduce waste, 19 per cent made an attempt to reuse and 25 per cent increased the amount they recycled after considering the campaign's message. However the surveyors acknowledged that these figures are claimed rather than demonstrated changes in behaviour. Overall the conclusion of the evaluations was that such awareness campaigns need to be on-going in order to maintain attention to the issues involved and that more detailed evaluation mechanisms need to be developed to track the effectiveness of different media interventions. It was also felt that awareness initiatives in the future should be devolved to sub-national scales. Money has subsequently been made available for regional awareness programmes while the Department for Environment provides co-funding for environmental awareness projects at the local level through Local Agenda 21 Environmental Partnership Fund (previously the Local Environmental Partnership Fund). The amount of funds available in these sub-national schemes are, however, rather smaller than the original budget for the *Race Against Waste* campaign.

In 2004 the EPA produced a review of waste management policy changes in a document entitled *Waste management – Taking Stock and Moving Forward* (EPA 2004). While noting the enduring commitment to the concept of integrated waste management and the EU's waste hierarchy the review identified a move away from a system of individual local authorities as sole providers of waste services towards regionalization and greater private sector participation. As a result of these changes attention to the movement of waste across administrative boundaries, waste prevention, markets for recyclables, biodegradable waste and greater producer responsibility was mooted. Most significantly the review reiterated a commitment to charging for waste collections based on usage. Waste charging in this way was seen as being both an equitable and effective incentive in the diversion of waste from landfill disposal. Local authorities were advised through a government circular to ensure that permits for waste collection included a condition to have a pay by weight or pay by volume mechanism in place by 1st January 2005 (DoEHLG 2004; CIR WIR 04/05). As long as the fundamental principle of use based charging was adhered to local authorities were given discretion as to which system (paying by weight or volume for example) they wished to develop. Recent analyses indicate that local authorities with pay by use systems in place have experienced reductions in waste being presented for collection (O'Callaghan-Platt and Davies 2006). There is concern however that a liberal interpretation of pay by use, based simply on whether a household chooses to pay annually for a large bin or a small one for example, has reduced the overall impact of the mechanism to date.

From the description of waste policy detailed above, and summarized in Table 5.3, it is apparent that waste management policy in Ireland has undergone significant changes that mirror those within environmental policy generally. There has been a clear attempt, for example, to move from a process of implementing incremental changes to disparate pieces of legislation towards a more comprehensive and integrated waste management system that includes a significant phase of planning.

More specifically there has been an acceptance, in central government at least, of incineration and recycling as key mechanisms to divert waste away from its traditional disposal to landfill in order to meet EU Directives. Yet incineration facilities are still not operational and recycling figures, while improving, are not currently sufficient to satisfy either the EU or those who call for a more dramatic movement away from disposal modes of governance to diversion and prevention. There is then a rhetorical commitment to reducing the environmental impacts of waste, but it is still primarily economic efficiency and public health discourses that predominate in legislation. These movements are reflected in the diversification in waste management mechanisms that have been developed. Simple, weekly single bin collections of unseparated waste to be disposed of at the town dump are becoming less common and a much more complex and variable picture of separated kerbside collections, bring centres, civic amenity sites is developing. There has also been an embracing of fiscal mechanisms to drive waste management behaviour as embodied in the plastic bag and landfill levies and more recently in the introduction of pay-by-use waste charging systems. Targets for diversion from landfill and for recycling have been made explicit and national educational campaigns for improving household waste behaviour have emerged.

These changes have not taken place within a vacuum and they are the result of complex interactions between a range of actors and agencies operating at a variety of scales in attempts to influence systems of governance. The influence of different scales of governing is of paramount importance in the geography of waste governance in Ireland. Local authorities were the primary actor in waste management until the late 1980s when the influence of the EU, primarily through Directives, became more significant. These Directives demanded that member states complied with greater requirements for waste planning and set in motion processes to divert waste from landfill. In turn the national government developed a framework for local authorities to implement these demands; a framework that was modified when local authorities failed to adopt the required structures for waste management that the national government desired. However a simple picture of a linear, top-down process from the supra-national to the local obscures a range of issues that affected the interactions between these scales of government and ignores the impact of non-governmental actors in shaping waste management policies and processes.

The following section considers the interactions between the multiscalar actors and agencies; the relationships between various tiers and spheres of waste governance. It focuses specifically on how the governing rationalities, that are embedded within the waste policy initiatives and reflected in governing mechanisms, have emerged.

Table 5.3 Irish waste policy interventions

	Title	Date	Key Details
Policy Documents	Changing Our Ways	1998	• Targets for waste management • 50% of household waste and 65% of biodegradable waste from landfill by 2013 • 35% of municipal waste and 85% of construction and demolition waste to be recycled by 2013
	Preventing and Recycling Waste: delivering change	2002	• Proposing recycling infrastructure improvements • Mooting producer responsibility initiatives
	Waste Management: Taking Stock and Moving Forward	2004	• Regionalization • Private sector service provision • User-based charges to be established by January 2005
Policy Initiatives	National Waste Prevention Programme	2004	• Waste audits • Advice and demonstration projects • Grant assistance
	National Waste Awareness Programme (Race Against Waste)	2003	• Business and household focus • Mass media campaign • Website information
	Cleaner Greener Production Programme (EPA)	2001	• To improve environmental performance, particularly small and medium enterprises • Awareness-raising • Technical training and financial incentive mechanisms
Legislation	Waste Management Act	1996	• Permits for commercial waste activities and enforcement powers • Monitoring and data collection • Waste management planning (hazardous and municipal) • Obligation to have due regard to waste prevention and recovery

	Waste Management (Amendment) Act	2001	• Plastic Bag Levy • Landfill Levy • Responsibility for adopting plans removed from elected officials and reallocated to County/city Manager
	Protection of the Environment Act	2003	• Enhancing enforcement provisions in waste field • Updating planning, licensing and permitting procedures • Reallocating responsibility for review, variation and replacement of waste plans from local councillors to County/City Manager
Policy Instruments	Plastic Bag Levy	2001	• Reduce consumption of plastic bags by targeting consumer behaviour • Levy of 15c per standard plastic bag (rising to 22c in 2007) imposed at point of sale
	Landfill Levy	2002	• To address externalities associated with landfilling waste • To address obligations under EU Landfill Directive (99/31/EC) • Levy of 15 € per tonne
	National Waste Database (three year cycle until 2002 then annual reports)	1995	• Data on municipal waste generation, recovery and disposal • Packaging waste recycling • The export of waste, including hazardous waste • Municipal waste landfill, and other waste, infrastructure
	Producer Responsibility Schemes	various	• Packaging (legally binding) • WEEE (legally binding) • Farm plastic (legally binding) • Construction and demolition waste (voluntary)
Funding Schemes	Environment Fund	2001	• To develop recycling infrastructure from proceeds of plastic bag and landfill levies • To develop public private partnerships • To support the National Waste Prevention Programme

Interactions

While the previous section identifies the importance of national policy interventions from Government as structuring forces in the processes and practice of waste management it was also recognized that these interventions are not developed in isolation from other spheres of governance or tiers of government. In this section attention will be paid to the other scales of government that are influencing the shape of waste policy in Ireland and to the private sector and civil society.

The EU is a particularly important actor in terms of dictating the overarching goals of waste management policy within member states although there remains a level of discretion available to member states in terms of how those broad goals are achieved. However it is processes of participation and exclusion, negotiation and resistance that go on behind the scenes, in lobbying campaigns or in the popular media, that are of interest here. To put this another way it is attention to the regimes of practice, the interactions between tiers and spheres of governance that are of concern. For while it is clear that waste governance exists, that is that waste is governed by actors beyond formal government, it is not clear from policy statements and documents how the various actors at particular scales or from different spheres of governance actually interact.

As was highlighted in Chapter 4 there is a history and persistence of negotiated compliance within environmental governance regimes in Ireland and the waste arena is no exception. For while policies and programmes undoubtedly have an important role in shaping waste landscapes the ownership of the collection and treatment of waste is also significant in terms of having control of the waste stream and developing waste infrastructure. Until recently local authorities were the primary providers of municipal waste services either in terms of delivery or infrastructure provision and the private sector waste industry was characterized by small-scale localized waste collectors who worked mainly for commercial and industrial operations. Since the publication of the *Changing Our Ways* policy statement in 1998 there has been increased support from central government for greater private sector involvement in a wider range of waste management services. The private sector has responded positively to this support as rising landfill prices and limited disposal capacity have stimulated commercial development of waste disposal and recovery.

A report by Forfás (2006) noted that, in contrast to many European and industrialized countries where the public sector controls municipal waste collection while industrial and hazardous waste is primarily collected and treated by the private sector, Ireland's municipal waste stream is predominantly privatized. Only 40 per cent of municipal waste is publicly controlled and the rest is managed either through the local authority sub-contracting out services to the private sector (around 10 per cent) or through purely private collection (just under 50 per cent). The most unusual feature of this arrangement is the large amount of municipal waste collection without any municipal involvement in establishing a contract or determining the flow of waste. Although the public sector still dominates in larger urban areas such as Dublin, Galway, Cork and Waterford there is no longer any public control of the waste collection in some local authorities. In the Forfás benchmarking study only New Zealand and Ireland out of the case study countries (which also included

Denmark, Sweden, Scotland, the Netherlands and Singapore) exhibited such purely private collection of municipal waste and even in New Zealand the level of purely private collection was much lower at only 10 per cent.[4]

Central government agencies and departments support this privatization, but concerns have been voiced about uncertainty in terms of the respective roles and responsibilities of public and private actors in delivering ISWM. The EPA in particular calls for plans to be updated to ensure 'effective engagement with the private waste industry' (2004, 38). Allied to this call is recognition that there is no direct regulator overseeing the liberalization of the waste sector in contrast to energy and telecommunications. In the absence of legislative requirements from the EU to establish a waste regulator the EPA calls for the structure and operation of the waste market to be kept under close scrutiny to avoid anti-competitive behaviour. This is particularly important given the lack of tendering procedures in many cases where waste collection processes have been transferred from local authorities to the private sector. In 2005, the Irish Competition Authority expressed its view that the predominance of purely private collections do not work well for the consumer and concluded that competitive tendering, rather than price regulation, was the most appropriate means of ensuring both good quality service and best value for money. In addition to monitoring competition the EPA has also established an Office of Environmental Enforcement which is focusing particularly on the waste sector and a programme to enforce the waste code is being funded through the Environment Fund in order to support more visible and frequent inspections of waste activities. These enforcement agents are supported by the Protection of the Environment Act 2003 that gives them greater powers and the ability to levy higher penalties. In 2006, following the government White Paper *Regulating Better*, a consultation document on the regulation of the waste management sector was launched in order to solicit views on the need and form of regulating for waste (DoEHLG 2006b). The privatization of collection and disposal is a central feature of waste management practices and already IBEC, the Irish Business and Employers Federation, has voiced its objection to the formation of as waste regulator. However the private sector has been pivotal in shaping the very foundations of waste management planning in Ireland in other ways, through the activities of environmental consultancies in developing waste plans and waste awareness campaigns. The role of one, now multinational, consultancy is particularly significant. Not only did the organization draft 80 per cent of the waste management plans nationwide, they were also successful in their tender to run the national waste awareness campaign *Race Against Waste* and the *Dublin Waste-to-Energy* information initiative. The consultancy's vision for waste is one of integrated waste management including, most controversially, incineration. The waste management plans produced, the awareness campaigns constructed and the informational initiatives conducted have all remained faithful to this vision. The success of the consultancy given the obvious similarity between its plan for waste management and that of business and central government is perhaps unsurprising, but it is unusual for one organization to have such a central role in the decision making arena of an economically developed country.

---

4   Competitive tendering for waste collection is however common in other countries.

Together these factors ensure that the private sector is a powerful actor in Ireland's waste management landscape. Private sector interests are engaged in waste debates through the consultants who drafted the waste management plans, via the private sector companies that collect and dispose of waste throughout the country and through the influence of business and industry lobby groups in policy discussions. The role of civil society actors in waste matters is less clear from the simple analysis of policy developments described above. While community groups did participate in the consultation phases of waste planning, and although community based resource organizations do exist, the numbers involved in both cases are small (Davies 2003; 2007). Still, even where more significant involvement of civil society actors has occurred, such as through the anti-bin tax protests in Dublin in 2003 or within the numerous community campaigns against proposed incinerators, their participation is rarely acknowledged in policy statements.

In analyses of environmental governance, including waste (see Boyle 2002; Bulkeley et al. 2005; Davoudi 2000), it is the inter-relationship of three sectors of society – public, private and civil – that fundamentally shapes outcomes. However most attention to waste governance in Ireland has focused on scalar interactions within the public sector rather than the intersections between different spheres of governance activity. Although there is an emerging body of literature on individual and household attitudes and actions in relation to waste (see Davies et al. 2005), and some recent consideration of voluntary packaging agreements with industry (Cunningham and Clinch 2005), attention to the role of civil society in waste management is absent from formal policy documents.

Even when civil society actors do gain media or political attention they tend to be characterized simplistically as self-interested NIMBYs, anti-development luddites or political opportunists (Davies 2003). This apparent marginalization of the civil sphere is surprising given the Irish Government's recognition that sustainable development (including waste management) requires the participation of all sections of society (Comhar 2002; DoELG 1997). Equally it seems to ignore the positive impact that civil society organizations have had on waste management overseas (Liss 2001; Luckin and Sharp 2003; ZeroWaste NZ 2003) and the influence that civil society organizations have made in other areas of social life in Ireland (Connolly 1997; O'Donovan and Ward 1999).

There is a growing body of research that claims certain actions of civil society in the waste field, such as community based recycling organizations (CBROs), produce 'benevolent effects' for society (see Foley and Edwards 1996) and for the environment (du Preez and White 2005; Luckin and Sharp 2003). However only a handful of such operations exist in the Republic in comparison with over 350 such organizations in the UK and a similar number in New Zealand. Research conducted with three arenas of waste related civil society action – CBROs, protests against waste charges and anti-incineration campaigns – revealed a range of constraints on activities that, at least partially, account for the current marginalization of civil society from policy making circles (Davies 2007; forthcoming). Although there were differences within and between the areas of civil society action four common factors were identified as the most debilitating: a lack of funding and resources; low status; limited access to policy making environments; isolation from other elements of civil society. These factors interact to compound the conditions of constraint in which waste-related civil society groups operate making it difficult for them to 'accumulate significant power to alter the dominant governance cultures in which they find themselves' (González and Healy 2005).

The civil society activists all identified both human and financial resource problems as a restrictive force on their operations. For CBROs the problem was seen as a lack of resources available either from the state, the private sector or charitable organizations to support their activities on a daily basis. To compound matters even when finance was available the conditions attached to the money were often felt to be onerous. For the anti-bin tax and anti-incineration campaigns the problem was more likely to be in relation to either raising funds for marches, demonstrations and newsletters or supporting people through legal challenges. In all cases the groups felt that this lack of financial clout contributed to the perception in public and private sectors that the operations or activities of waste-related civil society were of little significance.

The difficulty civil society groups encountered in attempting to communicate the contribution they were making (or could make) to society and waste management compounded feelings of insignificance within decision making environments. Again this manifested itself in different ways between the three areas. Amongst the CBROs there was a view that government was wedded to simple economic cost effectiveness in managing waste and that the added social, economic and environmental benefits of their activities (providing jobs to disadvantaged communities for example) were not appreciated. Within the anti-bin tax campaign the protesters clearly articulated their concern for political empowerment of marginalized low income groups within the city but the identification of the protests with predominantly left-wing political groups meant that their broader messages about justice and equity could be dismissed by the larger, more mainstream parties as political opportunism. In the same way anti-incineration campaigners have struggled to situate their concerns as both local issues for communities and as matters of national (and international) interest.

Without a clear line of access to policy making or a voice in policy deliberations it is unsurprising that waste-related civil society groups articulate a sense of isolation as a barrier to the effectiveness of their activities, but this was accentuated by a lack of co-ordination between local organizations. The lack of a nationwide umbrella lobby organization for CBROs, for example, was of concern to those who felt that they could benefit from scaling-up their voice in policy circles and from being better engaged with similar organizations around the country to disseminate good practice and share experiences. This concern was less dominant amongst the anti-bin tax campaigners because they were focused very much on grassroots communities, in some cases even streets, but at the same time many of the community campaigns benefited (at least in an organizational sense) from the participation of national (if small) political parties. Despite this the anti-bin tax protests failed to link up with CBROs or anti-incineration campaigns even though there is common ground between the groups in terms of equity and justice concerns in relation to waste management. Some interviewees suggested that this isolation was the result of a narrowness of vision amongst some of the political activists in the bin tax campaigns who focused on structural economic issues (taxation) to the exclusion of wider societal concerns, but there are other possible explanations. Parochialism in environmental campaigns has been identified by a number of researchers in Ireland and certainly there is a weakly organized and networked national environmental lobby (Allen 2004; Leonard 2006).

The above discussion indicates that policy interventions are the result of interactions between different actors and agencies operating at a variety of scales from the local to the supra-national, but what impact have these interactions had on waste management? The following section examines the outcomes of interactions between tiers and spheres of waste governance in Ireland.

Outcomes

One problem when attempting to evaluate the outcomes of waste governance in Ireland is the lack of accurate and consistent information on waste generation and management practices. A variety of approaches to data collection have been undertaken in the development of waste management plans and different frameworks have been adopted for projecting future waste arisings. Some plans including figures in absolute terms while others expressed future scenarios in terms of rates of increase over a specified base year. The same is true of recycling and recovery figures and forecasts. The result is a variegated body of information from which it is difficult to produce a national summary of waste statistics. Nonetheless the EPA has estimated that there has been a progressive and significant increase in municipal waste arisings since 1995 due to rapid economic and population growth (8 per cent between 1996-2002 and 4 per cent in 2004) and declining household size (EPA 2004). It is also suggested that the better monitoring and measurement of waste volumes means that a more accurate picture of actual waste figures is now visible and that previous statistics were underestimating the amount of waste being produced. Such was the underestimation that in 2001 volumes of waste being produced had already surpassed medium term projections in many of the waste plans. Overall there are two distinct camps when it comes to evaluating the outcomes of waste governance. First the optimistic view, primarily held by Government, is presented and this is followed by the more critical readings of progress.

Good Progress?

Despite the problems with attaining accurate data the EPA produced a national overview of waste management plans in 2004 that reflected on the progress made on the objectives of the waste management plans up until the end of 2003. It reported that at a national level significant progress had been made on the rolling-out of segregated collection of dry recyclables so that almost 42 per cent of all households received such collections in 2003 compared to around 5 per cent in 1998. In terms of basic recycling facilities (called bring banks in Ireland) there are now more than double the amount of sites compared to 1998 with a national density of around 1 per 2,300 people. Similarly civic amenity sites have risen from 30 to around 55 (and continue to rise) and most accept an extended range of materials. In the last few years local authorities have indicated at least a 25 per cent rise in the volume of waste being accepted at 'bring banks' and civic amenity sites (EPA 2004, 16).

Overall a combination of improved facilities and collections have led to increases in the recovery of municipal solid waste from 8 per cent in 1995 to 13 per cent in 2001 and 34 per cent in 2005. The EPA is convinced that this momentum will be sustained as the full impact of government funding for waste infrastructure becomes visible and feels 'there can be reasonable confidence that further significant progress will materialize in the short term under many of the recycling infrastructure headings' (EPA 2004, 17).

Contrasting Interpretations

Forfás, Ireland's national policy and advisory board for enterprise, trade, science, technology and innovation, provides a less positive reading of the current waste situation. In 2006 they produced a report that predicted significant problems ahead for waste management in Ireland given the shortfall between the investment target for waste management infrastructure funding detailed in the National Development Plan (NDP) of €825 million (which includes €571 million private investment) and the actual funding which has been estimated at €250 million and has come predominantly from private investment. Beyond this it is reported that

> the level of investment in the current National Development Plan was not sufficient to provide the level of infrastructure envisaged in the regional waste management plans and estimated that it will require a minimum investment of €2 billion to deliver the main elements of these plans (Forfás 2006, 7).

The same report also makes clear that while recycling rates are increasing Ireland still has a high municipal waste generation per capita figure at 777kg compared to other countries against which Ireland's waste management performance has been benchmarked (Singapore, Denmark, the Netherlands, Scotland, Sweden, Austria and the Czech Republic). Although the Irish figures may be inflated because of different definitions of municipal waste across the countries investigated the high trends reflect the economic growth of the Republic and a failure to completely decouple waste generation from that growth.

It is also clear that while recovery rates have improved this has been mainly through increases in the recycling of packaging waste. Diversion of household waste stands at 19 per cent, far below the national target of 50 per cent by 2013. By 2006 Member States were restricted to landfilling a maximum of 75 per cent of the total weight of biodegradable material generated in 1995 (the baseline year), however in 2004 Ireland was landfilling 101 per cent of biodegradable waste based on 1995 levels. Even though Ireland negotiated a four year derogation on the implementation of the EU Landfill Directive, because of its heavy reliance on landfill, this will still be a hard target to meet without radical developments in landfill diversion. Unsurprisingly then landfill capacity remains a critical issue with an estimated average of eight years remaining nationwide with Dublin and Donegal each having less than five years left (EPA 2005).

Concern about capacity for managing waste is not restricted to availability of landfill as noted by Forfás (2006). In 2005 the Republic was exporting waste with 30 per cent of municipal and 70 per cent of hazardous waste being disposed of outside national boundaries and there was only one plastic, one paper and one glass recycling/reprocessing facility in operation in Ireland. Other countries, such as Scotland, New Zealand and Denmark, that are comparable in terms of population and waste generation have more diverse and extensive indigenous recycling/ reprocessing landscapes. This level of exporting has arisen because it has been seen as the most practical (given the lack of indigenous capacity) and cost effective solution, however these benefits are dependent on relatively cheap transport costs and destinations willing to accept exported waste. In 2005 Germany increased its incineration gate fees and ceased accepting waste for its incinerators from overseas (including Ireland) when its capacity was reduced due to extra waste being diverted from landfill to meet its own EU Landfill Directive targets (EPA 2005). Overall Forfás (2006) conclude that it is Ireland's limited access to waste treatment solutions that will constrain its ability to achieve the targets for diversion from landfill set down at the EU level. In particular the report emphasizes the lack of incineration facilities as a significant problem for waste management capacity. The commitment to incineration may, of course, be linked to the fact that the authors of the report are RPS-MCOS the same consultancy that drafted many of the waste management plans for Ireland and whose plans all contained provision for incineration facilities.

In the North East Waste Planning Region planning permission and a waste licence have been granted for an incinerator although negotiations for an extension of the facility will require new permits and the developer involved in the case has recently suggested that they will not begin the construction of the incinerator until access to landfill is constrained in Ireland. The managing director of Indaver Ireland is quoted as saying that

> for an integrated waste management system to be successful, landfill capacity must be restricted, landfill bans on recyclable and combustible waste must be imposed and landfill taxes must be increased to ensure the viability of recycling and waste-to-energy infrastructure (Jennings 2006, 5).

In Dublin the procurement process for an incinerator is complete and a company has been awarded the contract to design, build and operate the incinerator. The planning process is underway and applications have been made to An Bord Pleanála for planning approval and to the EPA for a waste license. However in other areas no significant progress has been made and significant local resistance still exists. For while a representative of the Irish Government made a statement at the 2006 International Solid Waste Association Conference suggesting that the debate about incinerators was over in Ireland, as that the activists had been convinced there was no danger from dioxins, opponents to an incinerator in Dublin have already lodged over 3000 objections (Kelly 2006). In addition there remains vocal opposition to the incinerator for the north east region, due to be built in Carranstown, in terms of its potential impact on heritage, landscape and water supplies and growing concerns from anti-incineration campaigners that the government has left themselves a hostage of incineration companies (Davies, forthcoming). Pat O'Brien from the No Incineration Alliance campaign in the north east stated that

> Indaver came here telling us we had to have incineration immediately because we had insufficient landfill supply. Now they are trying to tell us we have too much landfill available and they are actually trying to get our government to tax existing and futuristic landfill sites to help make there planned expensive out of date, incinerator, financially viable (in Jennings, 2006, 5).

There was also concern that Indaver have managed to get a condition of the waste license loosened so that they are able to accept waste from outside the waste planning region. Opponents now fear that the proposed incinerator will effectively become a national, even international, facility.

In terms of the foundational vision for waste management in Ireland the EPA and central government remain committed to the concept of integrated waste management that is based on the waste hierarchy of waste prevention, followed by minimization, re-use, recycling, energy recovery and finally sustainable disposal of residual waste. In reality this policy commitment has been translated into the identification of what has been called maximum achievable levels of recycling and biological treatment, which led to the judgement that incineration facilities are necessary to meet EU targets for landfill diversion at a national level. Critics of this approach have suggested that the practice of integrated waste management has been too preoccupied with the lower end of the hierarchy rather than prevention, minimization or re-use and that maximum achievable levels depend not just on technical feasibility but also on political desire to support incineration. There have been calls from environmental groups and some opposition political parties to adopt a zero waste policy, but the government has roundly dismissed these suggestions on two grounds. The first claims that zero waste policy is effectively the same as integrated waste management, despite the fact that zero waste advocates would not support incineration, and the second suggests that zero waste is idealistic and unachievable. The EPA stated that 'no country has shown the "Zero Waste" aspiration in its purest sense to be an achievable objective. Indeed, even in the limited number of localized situations where "Zero Waste" has been pursued, it has still not been proven as an effective approach to waste management' (2004, 22). While the voices of the zero waste movement, predominantly emerging from the civil society sector, in Ireland have not been able to influence the discourses of waste management in government circles there has been some recognition of the need to move towards waste prevention and minimization activities. Interlinked with its Presidency of the EU in 2004 Ireland launched the National Waste Prevention Programme 2004-2008 (NWPP), led by the EPA and supported by the Environment Fund. While this programme recognizes the importance of waste prevention it is interesting to note that a primary stated aim was to 'improve data on waste arisings so that a sound basis for measuring and monitoring of the programme's impact can be established' (EPA 2004, 29). Critics have argued that such a focus on data collection allows for business as usual and shies away from the more politically contentious aspects of trying to reduce the amount of waste being produced by different sectors and in specific waste streams. They also pointed out that many of the developments outlined in the NWPP had been identified in 2002 waste policy document *Preventing and Recycling Waste: Delivering Change* (DoELG 2002) with no explanation for the delay in implementation (Green Party 2004). The same kind of slippage has occurred with the formation of a Recycling Consultative Forum identified first in 2002, then to be established by the end of 2004 (EPA 2004) and then by the end of 2005 (Department of the Taoseach 2005). While the EPA remains convinced that the 'vanguard of national waste policy is perceptibly shifting towards prevention' (EPA 2005, ix), it also calls for more supra-national level direction in this area. So while waste planning has been identified as a process for regions, waste prevention is perceived as a practice that requires transnational action, which brings back attention to the scalar politics of waste management.

To summarize, both private sector and civil society organizations have expressed concerns about the ability of current waste governance systems to operate effectively, although their areas of concern differ. Business and industry tend to be worried about the impact that slow infrastructure development (in particular in relation to incineration), continued increases in waste generation and a dependence on exporting waste will have on further economic expansion and competition (Forfás 2006). Civil society actors, and environmental groups in particular, focus more on the need for further demand management and greater attention to waste prevention and minimization. Both the private and civil society sectors have expressed concerns about the level at which waste management decisions are being made which suggests a scalar politics is important within waste governance practices.

Scalar Politics

It has been established that a key feature of the waste management planning process since the 1996 Act has been the support from central government for a regionalization of the planning units that deal with waste management. This is a change from previous practice that operated on a county or city basis. It was in *Changing Our Ways* (DoELG 1998) that encouragement was explicitly articulated in order to 'secure more efficient provision of infrastructure and services' (EPA 2004, 23). Notions of efficiency here are based on the assumption that economies of scale operate in terms of waste management such that regions provide a viable framework in planning and volume terms for the development of the integrated solutions advocated in the waste management plans. In addition, and significantly from a governance perspective, the EPA claims that a regionalized structure of waste planning also facilitates 'innovative solutions and a favourable climate for the creation of beneficial partnership arrangements between local authorities and the private sector' (EPA 2004, 23).

Another justification for regionalization is that regions provide the best solution to the challenge of the proximity principle. This argument is made on the grounds that a balance has to be struck between optimising costs and efficiencies of waste management facilities and minimising environmental damage through energy and pollution costs of transporting waste. Yet the issue of adhering to the proximity principle is somewhat clouded by the possibility of transporting waste from one planning region to the next. The EPA suggests that to restrict planning permissions for certain waste infrastructures so that they can only deal with waste arising within the area to which the waste management plan applies is 'too blunt an instrument' (EPA 2004, 25) and that the inter-regional movement and treatment of waste should be provided for in certain circumstances. In May 2005 Minister for the Environment Dick Roche gave policy directions on the movement of waste. This direction stated that the application of the proximity principle in the context of waste management does not entail interpreting administrative waste boundaries in such a manner as to inhibit the development of infrastructure that would support the attainment of national waste policy objectives. Just to reiterate, central government has explicitly encouraged the regionalization of waste planning, yet in the context of transporting waste it seems that national policy objectives hold sway.

The legal movement of waste across administrative boundaries is not the only contentious aspect of transporting waste. Illegal dumping has also received significant attention in recent years following high profile cases where waste has been transferred illegitimately across borders both to Northern Ireland and overseas. In 2004 Dutch and Belgian authorities discovered shipments of waste that originated in Ireland and were incorrectly listed as recyclables. Under EU Regulation 259.93 on the Transfrontier Shipment of Waste exports of waste have to obtain consent (with certain exceptions) from the relevant competent authorities in countries of transit and destination. Only 'green list' waste – including uncontaminated segregated materials such as waste paper, cardboard and glass – do not require prior notification. Under these regulations the movement of improperly notified waste is illegal. In one case the illegal waste originated from nine counties and nine different waste companies. The case is further complicated in that waste brokers are frequently used to transfer waste overseas. In 2005 Dick Roche the Minister for the Environment established a policy Direction on Section 60 of the 1996 (Amended) Waste Management Act to encourage an intensification of action against illegal waste activity and enhance the response of local authorities and the EPA in ensuring the protection of the environment and human health and the prosecution of offenders (DoELG, 2005 Circular WIR: 04/05).

These events indicate that there are evident struggles over scale and influence in relation to waste in Ireland. Local authorities were identified as the lead institutions for municipal waste planning, albeit with caveats for national government recovering some central control over local plans if desired. However most local authorities, following encouragement by national government, opted to collaborate with other local authorities on their waste plans, despite the weak regional structure of government in Ireland (Laffen 1996; O'Leary 2003). As mentioned earlier the regions that emerged for waste management were created solely for waste planning purposes and have no other statutory roles or responsibilities, indeed they do not fit with the regions identified in the National Development Plan (DoELG 2000). Decisions, such as these, to locate policy making at particular scales are inherently political as they can have significant impacts on the definitions of issues and on the potential resolutions to problems (Boyle 2002). For example, the regionalization of waste plans, through economies of scale, creates sufficient markets for large waste management facilities, such as incinerators and large engineered landfills, to become cost-effective. Equally not opting for regionalization in waste management planning might also rule out certain processes through a lack of economies of scale. In Wicklow, which chose not to join with other counties and produced a county plan, for instance the plan notes 'thermal treatment at a state of the art centralized facility serving county Wicklow alone is not currently economically feasible' (Wicklow Council 2000, 12).

While it could be argued that the move to regions in waste management is a positive attempt by government to decentralize decision-making it creates problems for co-ordinating cohesive restructuring strategies for industrial production processes which might encourage minimization and prevention of waste at source and it also makes establishing a coherent recycling infrastructure extremely difficult. This lack of national framework has been identified as a deliberate scalar strategy designed to simultaneously avoid having to deal with potentially controversial developments at the national level while locating responsibility to 'weak, fabricated regional groups' (Boyle 2002, 185).

Despite the concerted effort by central government to impose particular waste management strategies, by effectively removing channels through which civil society actors might voicing their opposition, there still remains potential for resistance to waste management policies, through creative and extended use of network alliances and coalitions. As noted by Darier (1999), following the ideas of Foucault, there are rarely situations of absolute domination and there are generally some opportunities for resistance. The relatively closed political structures in Ireland where participatory channels are being increasingly limited in relation to waste management could lead campaigners to seek more robust international linkages to press their claims more widely. This would mean not just using international contacts as information sources, but also utilising their potential as a collective power base from which pressure can be placed on the Irish government. In conjunction with the development of new virtual political spaces and consolidated international strategies, civil society groupings might seek to strengthen their national profile, creating stronger alliances with other national social organizations, powerful economic interests and lobby groups. Here lessons could be learnt from movements in other countries, which have experienced similar challenges albeit in differing political contexts (Davies 2004).

It appears that waste management in Ireland is defined by strongly hierarchical relations in terms of governmental influence from the top down; from the EU to national government to local government. However the picture is complicated by the strength and influence of private sector interests both in the formation of policy structures through defining visions for waste plans and through the delivery of waste collection and disposal services. The top-down, or trickledown, model of governing also needs to be modified to account for the behaviour of national government in the waste field. It appears that while the national government is happy to see waste plans, educational initiatives and collection and disposal services defined by the private sector it is more reluctant to devolve decision making powers to the local level. These institutional relations are best illustrated through in-depth analysis of particular moments in the evolution of waste management policy and no other moment has been so controversial as the development of waste management plans. The following section reviews the interaction of a range of actors from public, private and civil society spheres of governance who came together in one context in the light of the development of the Connaught Regional Waste Plan located on the west coast of Ireland (see Figure 5.1). It was a process that was replicated in many locations around the country and the responses that emerged were influenced by both local circumstances and actors far removed from the waste planning region such as multinational waste companies and transnational advocacy networks as well as interventions from national and European government.

The Waste Management Act and Incineration Politics in Galway

There are many interesting dimensions to the geographies of waste governance in Ireland, but the sanctioning of municipal solid waste (MSW) incinerators following the 1996 Waste Management Act has caused particular controversy. That Ireland currently does not have MSW incinerators makes it unusual within the EU (Greece is the only other country not to have MSW incinerators) and the proposal to construct them marked a sea-change in Irish waste management practice. Support for the MSW incineration facilities emanates primarily from central government and the private sector (that includes business interests as represented by IBEC and the waste consultants who drafted the plans and much of the waste industry), while opposition exists particularly although by no means exclusively, within the communities where those facilities have been proposed.[5]

That conflicts exist regarding the introduction of MSW incineration in Ireland is unsurprising given that similar conflicts, relating to large-scale incineration of both municipal and hazardous waste, have been occurring in the USA and Europe for over twenty years. These conflicts over waste incineration have been subjected to examination from a range of perspectives from geography to psychology and from economics to science and technology studies. Research has included examinations of technical processes of risk analysis and public perceptions of risk in relation to waste-to-energy incinerators, risk communication studies, psychological analyses of community responses to waste facilities and economic audits of cost implications for land value or through compensation packages (see Beder and Shortland 1992; Gray 1995; Kiel and McClain 1996; Lima 2004; Nieves et al. 1992; Petts 1992; Snary 2004; Zeiss 1998; Zeiss and Paddon 1992).

From a policy analysis perspective Dente et al. (1998) examine construction of incinerators across Europe, while Kuhn and Ballard (1998) similarly consider Canadian innovations in successfully siting hazardous waste management facilities. Perhaps the most well developed area of research in relation to conflicts over incineration has emerged within social movement studies, particularly with regards to the links between incineration conflicts and community activism. Walsh et al. (1997) extend traditional social movement theory in light of extensive analysis of grassroots challenges to waste incinerators, while Gerrard (1996) makes a direct connection between ideas of environmental justice and the siting of toxic waste facilities. One key development within this broadly sociological school has been

---

5   The aim of this case study is not to prove or disprove the case for (or against) incineration rather it is to examine how actors concerned with the governance of waste management have responded to conflicts surrounding the process.

the critical deconstruction of the NIMBYism (not in my backyard) concept, with researchers calling for a more sophisticated analysis of community opposition to waste incineration facilities (Fischel 2001; Hunter and Leyden 1995; Luloff et al. 1998). Linked to the deconstruction of NIMBYism has been a contextualization of conflicts through discourse analysis, and specifically the collective framing of anti-incineration discourses by activists (Kubal 1998). For the most part these studies of incineration conflicts focus on a single geographical site and the associated opponents and supporters of the incineration scheme. Studies have thus tended to be preoccupied with the locality, focusing on local communities, local governments and other agencies acting in the locale. Even when non-local aspects of incineration conflicts are acknowledged analyses are inclined to categorize them as simply exogenous resources waiting to be mobilized by local actors either for or against the facility under investigation (see Walsh et al. 1997). There has been little explicit or detailed consideration of how actors at different scales affect incineration conflicts. Adopting such a predominantly ageographical approach in the investigation of incineration conflicts has tended to isolate these debates from wider considerations of waste governance.

As detailed in previous sections incineration politics emerged in Ireland following the 1996 Waste Management Act and the requirement for waste management plans at either the county or regional level. Most of the plans that were produced by consultants as a result of this requirement included incineration as part of an integrated waste management strategy in order to meet the demands of the 1999 EU Landfill Directive. The only plans that did not contain an element of MSW incineration were those that were developed on a single county basis and even these only ruled out incineration based on an assessment of economic cost.

The waste plan for Connaught, as with the other plans produced across Ireland incorporated an integrated waste system that included targets for diversion from landfill such that nearly 33 per cent of waste would be thermally treated, and just over 48 per cent recycled by 2013. The plan indicated that the incinerator would be located within Galway City boundaries following the proximity principle that suggests facilities should be sited near to the centres producing most waste, although no specific sites were identified. It also proposed the rationalization of the existing landfills within Connaught into two major landfills one in the north and one in the south of the region. As the plan was being discussed by the local authorities involved a number of community groups in Galway were turning their attention to the issues it raised and coalitions of anti-landfill and anti-incineration groups were being forged.

The first opportunity for input from individuals and civil society in the waste management process occurred through the formal consultation mechanisms on draft management plans. The plans were placed on public access for a statutory two-month period required by the Waste Management Act. A low number of written submissions were made to the draft planning process throughout Ireland, which may seem to indicate little concern amongst individuals and organizations about the content of the waste plans. For example, during the draft consultation phase in the Connaught region (Galway City, Leitrim, Mayo, Roscommon, Sligo and Galway Council Councils) there were only five written submissions from the public or individuals and seven from NGOs compared to 12 from commercial groups and 18 from public representatives (Davies 2003). However this low level of participation did not reflect the extent of public concern about the waste management strategies being proposed and in 2000 Galway Safe Waste Alliance (GSWA) was formed as an umbrella group linking community groups concerned with the proposals of the waste plan.[6]

The remainder of this section examines the formation and evolution of the campaign against the proposed incinerator in Galway set against the backdrop of waste governance in Ireland. It draws on empirical research, conducted between 2003 and 2004, that involved two main strands: first, a content analysis of local and national newspaper coverage of waste management issues and incineration in the Galway area and second, in-depth qualitative interviews conducted with key actors in the incineration debates in Galway. The content analysis provided the background for the study and facilitated the identification of key actors, operating at different scales, for the interview phase of the research. The newspaper coverage provided textual detail on the kinds of arguments that were used by both pro and anti-incineration actors and follow-up interviews were conducted with actors drawn from a broad spectrum of backgrounds and perspectives that were supportive and opposed to incineration. These interviews included actors operating at different scales from the local community resident mobilized by the proposal to build an incinerator in their area to the international anti-incineration activist. Interviewees also included pro-incineration actors including the consultants who developed waste management plans and politicians involved in national debates for and against incineration. One methodological problem when studying issues that transcend scales and involve expansive networking is the large number of potential interviewees who may have participated in debates to various extents at particular times. The content analysis of the newspaper coverage partly overcame this problem, but in order to respond to the ebb and flow of participation a process of snowballing was adopted such that each interviewee was asked who else might be an important person to interview.

GSWA drew initially on national resources to support their anti-incineration campaign, using the work of the Waste Working Group (led by Friends of the Earth Ireland, Voice of Irish Concern for the Environment (VOICE) and other concerned individuals) to form the basis of their written opposition and seeking public support from national groups such as the Irish Farmers Association (IFA) and the Irish Doctors Environmental Association (IDEA). GSWA also participated in a loose national grouping called Zero Waste Ireland, formed in 2000 as an alliance of anti-dump, anti-incineration and local waste management groups. There was an international dimension to the anti-incineration campaign through communication

---

6   The groups include Galway for a Safe Environment, the most active group in the GSWA campaign, Galway Environmental Alliance (which includes An Taisce, Galway Cycling Campaign and Crann amongst others); Newbridge Action Committee; Ballinasloe Against the Superdump; Clontuskert Anti-Incineration Group; New Inn Anti-Dump Committee; Kilrickle Anti-Dump committee. While it may appear to be an unholy alliance between anti-landfill and anti-incineration positions, both coalesced over common agreement concerning the need for a zero-waste strategy.

with, and subsequently membership of, transnational advocacy networks (TAN) (Keck and Sikkink 1998) such as GAIA (Global Anti Incineration Alliance) and the Zero Waste Alliance. These networks provided a global dimension to arguments against incineration, linking GSWA with other community campaigns, and key international anti-incineration campaigners, across the world. Importantly the TANs GAIA and Zero Waste Alliance are not simply coalitions of opposition organizations; they are also proponents of alternative mechanisms for waste management, which conceptualizes waste as a 'resource in disguise' (Zero Waste Alliance 2004).

The use of ICT (information and communication technologies) by GSWA members and these other networks, in particular the design and use of web-pages[7] and e-mail list serves, facilitated the collection and dissemination of information about incineration between these globally networked organizations. GSWA developed arguments against incineration facilities drawing on the experience of other anti-incineration campaigns from across the globe and a case was constructed around the issues of need, risk and trust that: i) emphasized translocal concerns about the need to focus higher up the waste hierarchy to re-use, recycling, minimization and prevention ii) highlighted the uncertainty amongst the scientific community regarding the potential risks from incineration emissions and iii) questioned the ability of the national regulator (EPA) to effectively regulate the operation of waste facilities. Thus while there was a clear space of dependence for the anti-incineration campaigners created by the identification of Galway as the site for an incinerator, the space of engagement for the campaigners was much broader, both geographically through the TAN and conceptually, through the building of alliances with diverse interest groups.

As well as consulting nationally and internationally GSWA adopted diverse lobbying strategies at the local level making oral presentations to both Galway City and County Councils, organising public marches, collecting signatures for petitions and encouraging individual submissions to local authorities. Over 22,000 signatures, making up one in four of the population in Galway city, were collected through petitions and over two thousand individual submissions were made to local authorities. Public meetings peaked during 2000 with over 800 people attending one event, and a march against incineration in the same year drew over 2000 people to Galway city centre. The local support also generated financial resources that were used to fund the campaign and foster further participation in the transnational network. International anti-incineration and recycling experts, were flown in to Galway using these funds and members of GSWA visited landfill sites and incinerators in Germany and the UK. The networked connections of GSWA to other actors and organizations are summarized in Figure 5.2.

---

7 For example those hosted by Galway for a Safe Environment and Newbridge Action Committee http://www.go.to/gse and http://gofree.indigo.ie/~ljhannon/index.htm.

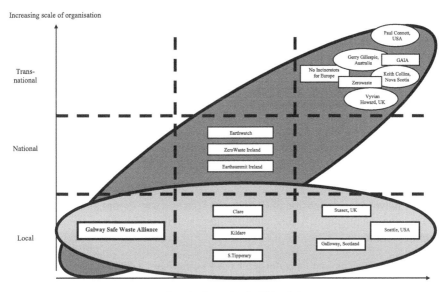

Figure 5.2 Networks of anti-incineration organizations and individuals related to Galway Safe Waste Alliance
Source: Adapted from Davies 2005.

While central government and private consultancies were disseminating positive messages about the need for incineration GSWA offered alternative sites of knowledge production, contestation and utilization (Escobar 1996). Campaigning approaches were shared and experiences consolidated amongst the networks of anti-incineration groupings. The spatialities and trajectories of resistance through networks were simultaneously local and transnational modes of political activity (Di Chiro 1997; Featherstone 2003). The international networking of GSWA enabled it to 'scale-up' (Perreault 2003, 12) their political voice, while consolidating collaboration amongst local communities that were the location of both political power and cultural reproduction. Pragmatically GSWA had to convince the local councillors that their constituents did not support incineration in order to persuade them to vote against the adoption of the plan at a full council meeting.

Key figures in the local campaign recognized the benefits of having strong resource mobilization capacity within the local community, as one member of the Galway Environmental Alliance (GEA) said

It was absolutely stunning, we had two engineers on the committee … incredibly experienced people … it's much easier to organize in a smaller community (GEA).

Of course the 'localness' of local resource capabilities is complex in an increasingly globalized world. Campaigners involved in the Galway campaign included those born and brought up in the region, those who had lived in the area for many years but who were originally from elsewhere either within Ireland or overseas, as well as those from the area originally, but who had moved away for considerable periods of time, returning only recently. GSWA brought together people with no campaigning background but who had areas of specific expertise, for example in relation to the technologies of incineration or the legal and planning system, with others who had long been involved in environmental struggles. As one member of the Galway Environmental Alliance said

[w]hen I heard about it [GSWA] I was thrilled, because I've been involved in alliances – I've lived in Toronto, Canada, before moving here and I've been involved in alliances and they were a very good way for a small group to wield a lot more influence by just combining forces (GEA).

The networking dimension of GSWA was seen as a way of maximising resources and avoiding duplication. As one member suggested the alliance was founded on the commitment to

sharing information and supporting each other's campaigns, it doesn't make sense for two groups to be working on the same issue, there aren't enough people. It worked both ways sharing support, raising awareness of actions and the like (GSWA).

In this way the campaigners clearly felt that the issue of incineration could not be defined in scale-exclusive terms such as local, nation or regional. Indeed the international campaigning expertise accessed through transnational networks was seen as invaluable for the strength of the case they were making to local councillors. A member of Galway for a Safe Environment (GSE) stated that

a lot of it [information] came from the GAIA network which is the Global Anti-Incineration Alliance ... we relied to a large extent on a number of experts who had fought against incineration internationally, like Paul Connett, Vyvian Howard, guys like that ... our general attitude was "look, there is evidence out there, in reputable places, that there are dangers" and then our other attitude was "there are other ways of doing this" (GSE).

While the expertise required to plan waste effectively, seen from the national government's point of view, was located solely with the engineering consultants in the formation of waste management plans, networking through physical meetings and virtual spaces (ICTs) enabled the anti-incineration groups to access information beyond the control of the consultants and government departments. There is resonance here with the view that

knowledge formation and power over knowledge in the global economy is moving out of control of the nation state, because innovation is globalized, because discourse and knowledge is outside the state's control, and because information is much more accessible than it was before thanks to technology and communications (Carnoy and Castells 2001, 11).

The evidence sourced from international experts allowed the groups to transcend the national governments labelling of publics and even local governments as without the sufficient knowledge to make waste management plans. Indeed members of GSWA explicitly commented on how crucial accessing information was to their campaign as they acknowledged that

> access to internet extended the information we could get hold of amazingly and the people who we could contact to ask for advice … we also felt connected, not alone … I couldn't have done the web-site, but the internet is out there for us all (GSWA).

However caution as to the empowering potential of web-networking is necessary as there remains a digital divide amongst the population and it is still a select minority who have the skills and resources to create and participate actively in the ICT arena (Pickerill 2000).

Despite the best efforts of GSWA central government and the consultants involved in drafting the waste plans, still characterized the anti-incineration campaigners as simple NIMBYists with an aversion to the location of waste facilities. The official Government response to the health fears raised by GSWA was to codify them as alarmist and without sound scientific evidence while at the same time providing assurances that science, engineering and technology could resolve any remaining technical difficulties with the processes of incineration. Undeterred by these reactions opposition reached a zenith in 2000 and both Galway City and County councillors voted unanimously to reject the Connaught Waste Management Plan as long as it contained a commitment to incineration. It appeared that the successful campaigning of groups in Galway, underpinned by the increasingly dense web of translocal and transnational anti-incineration groups, had convinced local councillors of the unacceptability of the Connaught Plan for communities in Galway. Yet, rejection of the plan did not mean an end to the incineration debate as threats of action against the Irish state due to non-compliance with the EU Landfill Directive were being discussed in the European Court of Justice.

In January 2001, with still no movement on the Connaught waste planning process the Minister for the Environment visited Galway and spoke out in support of incineration. Following this visit the councillors in both the city and the county were asked to vote again on the same waste plan. A second vote was held in Galway City in February 2001 and the councillors voted by an 8-7 majority for adoption. The councillors who changed their votes second time round were those affiliated to the ruling party coalition of Fianna Fail and the Progressive Democrats. The power of central government had the desired effect in Galway City, but Galway County Council were not to be similarly convinced and on their second vote, in April 2001, they rejected the plan again. Movement towards the adoption of the Connaught Plan was thwarted once more.

The county council in Galway was not alone in rejecting a waste plan. By the summer of 2001 there were still three local authorities within Ireland who refused to adopt the regional waste management plans presented to them. Rather than revisit the plans in order to see whether amendments could be made that were acceptable to the local councillors (and by association acceptable to the campaigners and the local communities) a decision was taken in central government to amend the 1996 Waste Management Act to resolve the deadlock. The 2001 Amendment Act incorporated a range of issues, but most pertinent to the waste management planning process was the proposal to relocate responsibility for the adoption of waste management plans from the local councillors to the county or city manager, a civil servant appointed by central government. Heated debates took place about this move within the formal structures of Irish Government (the Dáil, the Seanad and Select Committees), on the pages of newspapers and of course within those local authorities where plans were yet to be adopted.

Galway for a Safe Environment (GSE), a member of GSWA, were particularly vocal in their criticism of the amendment asserting that it was 'ignoring the will of the people' and 'attempting to impose waste incinerators on local authorities'. An article in the Irish Examiner in February 2001 quoted Dr Conchur O'Brádaigh, a key figure in GSE and GSWA, as saying that

> [t]hese proposals represent an admission of failure on the Minister's part. He has spent the last three years and up to £10 million of taxpayers' money attempting to convince local councillors that incineration is safe and sensible, including expensive trips to the continent. By taking away their decision making powers now, he is admitting that he has failed (Irish Examiner 2001, 10).

The political implications of this move were significant, particularly because only two years earlier the same government had passed the Local Government Act, which specifically enhanced the democratic powers of local government in Ireland. As one opposition TD stated

> [the Amendment] stands democracy on its head ... You cannot on the one hand, as in last June, put into the Constitution a provision which gives constitutional guarantee and protection to local government and then introduce legislation which removes from the democratic tier of local government the powers to make decisions on issues that affect local communities throughout the country (Gilmore, TD, Select Committee on Environment and Local Government 3/07/01).[8]

The response from central government was to characterize waste management planning as an issue of national interest and the opponents as an awkward 'local' minority not willing to make compromises for the common good and national development.

> The decisions of a small number of authorities have obstructed any prospect of progress by the majority and have thrown the planning process into disarray ... Our ongoing failure to respect our waste planning obligations has led to legal proceedings against Ireland in the European Court of Justice ... The Minister has to act in the *national interest* by taking steps that will facilitate the satisfactory completion of the planning process (Mr. Molloy, TD 3/07/01, Select Committee on Environment and Local Government, emphasis added).

---

8   Full transcriptions of debates conducted in the Dáil, Seanad and Select Committees are publicly accessible through the Government of Ireland web page at http://www.irlgov.ie.

The argument that the amendments were necessary because of a few intransigent counties holding the country to ransom was hotly contested

> [l]ocal authorities have not failed to adopt waste management plans, but have sometimes refused to adopt the plans of consultants (McCormack, TD Select Committee on Environment and Local Government 3/7/01).

This was the case in the Galway region, where both city and county councillors had voted to adopt a different waste management plan from that proposed by the consultants. This plan incorporated only Galway city and county rather than the six counties incorporated within the Connaught plan. Such a position would have negated the feasibility of thermal treatment through incineration within the Galway region, although it would still have required a landfill site, recycling and composting facilities all of which had been agreed by the councillors. So while central government claimed there was intransigence amongst the councillors, the councillors retorted that they were open to waste management planning, just not the plan they were being asked to vote on.

The concerns of opposition politicians to the amendments focused primarily on issues of scale and democracy, the tensions between local autonomy for waste decision making and the waste management crisis as a nationally significant issue. There was also concern however about the close relationship between national government and the private sector in terms of considering waste management options. Some suggested that there might have been an ulterior motive for the close relationship between the consultants and central government over incineration, encapsulated by one politician who said

> I do not understand why the Minister is so adamant in pursuing the incineration line. Is there some vested interest pushing that line? It is all very fine for consultants to say that incineration, within regions, is the solution to the problem. It is not a solution ... The only reason regions were proposed by consultants and adopted by the Minister was that there would be enough material for incinerators to keep them going (Mr. McCormack, TD Select Committee on Environment and Local Government, 3/07/01).

In one statement opponents, from whatever scale or perspective, to the waste management plans and the amendment of the act were dismissed by the then Minister for the Environment

> it is particularly galling to think that unnecessary fears are being raised and expert opinion is being ignored, credible statistics are being distorted, reports from the World Health Organization, European and UK sources are being discounted and rubbished ... we have an *alliance* of doctrinaire Greens ... who propose hopelessly impractical solutions to our waste problem ... we also have *globe trotting* self appointed 'experts' supplying a niche market with stock, simplistic answers, usually based on the experience of the 1960s and 1970s. We have political opportunists, who are happy to jump on any bandwagon they believe to be populist and that might get them elected. We have group clearly motivated by the *NIMBY* agenda (Noel Dempsey, Minster for the Environment 28/3/01, Seaned Debates, emphasis added).

Despite the raft of objections the amendment was passed and the making of a waste management plan became an executive (management) function. By giving the decision making power to centrally appointed civil servants the government was able to maintain its apparent distance from unpopular decisions over waste facility siting, while gaining more control over the decision makers.

By 2002 all waste management plans had been adopted and developments for MSW incineration are underway in the north-east and Dublin. Indaver Ireland, the Irish arm of the Belgium multinational Indaver, has been given planning permission and a waste license for an incinerator in Carranstown in Meath although an application to expand this facility requires amendments to these permits and Dublin city council has received an application for a MSW incineration facility in Ringsend within the city of Dublin (and over three thousand written objections). No further developments have been proposed in Galway to date and despite the considerable blow dealt to the GSWA campaign and to other anti-incineration movements in Ireland by the Amendment Act, there are still opportunities for creative resistance. Key individuals involved in GSWA are still working with GAIA and ZeroWaste and these activists remain quietly optimistic about preventing incineration in Galway

> I would say the public have forgotten about it [the incinerator] at this stage ... they'll say "ah we were going to get an incinerator a few years back – haven't heard anything about it in a while – I'd say it's gone" ... but there'd be the worse uproar they've ever seen, guaranteed, if the incinerator actually gets built (GSE).

Nonetheless the amendment of the 1996 Waste Management Act effectively pulled the campaigning rug from under the feet of GSWA and other anti-incineration campaign groups in Ireland. The initial success of the anti-incineration groups spatial strategy in generating internationally sourced arguments to present to local communities and their representatives (who at that time were responsible for deciding on the validity of the waste plan) was neutralized by the removal of decision making powers from elected officials and their relocation to civil servants. To put this in the terms of Cox (1998) the strategy of the anti-incineration groups to widen spaces of engagement, in terms of information gathering and campaigning tactics, but to focus the results of that consultation primarily on local councillors – the space of dependence – was fragile once central government enacted its powers to fundamentally amend policy making procedures. While there were significant local resources, financial and physical, within the Galway area, and key individuals championed international networking and the exchange of ideas with international expertise, they were all still volunteers trying to campaign against full-time politicians and consultants. The campaigning groups positions were at odds with the thrust of central government policy that was firmly aligned with the interests of large private sector organizations. Indeed relationships between the waste industry and waste legislators were further cemented in 2004 when a project manager from Indaver, the company planning to build two incinerators in Cork (hazardous waste) and Meath (municipal waste), was appointed as a Director of the EPA.

The analysis of GSWA located within a wider system of waste governance clearly indicates that activists adopt complex scalar strategies, and thus create intricate spaces of engagement, in their attempts to mitigate their problematic relation to a space of dependence. As such the politics of waste management are played out across scales rather than being confined within them and while resistance may occur in a place it is not necessarily localized. Networking, through TANs such as GAIA affords activists the opportunity to expand their political voices and bridge spatial distances in order to build coherent resistance positions. Yet, as demonstrated by the willingness of the Irish state to amend policy making procedures in order to silence pockets of resistance, even these transcalar strategies are vulnerable to the dynamism of political environments.

The case of incineration is a particularly fraught one in the Irish context, but similar patterns of marginalizing civil society actors and promoting market-led responses to waste management challenges can be seen in the lack of a CBRO sector in Ireland, the absence of civil society waste actors from policy making communities and in the treatment of protestors against waste charging in Dublin in 2003 (Davies 2007).

Conclusion

A key finding of this waste governance analysis of Ireland is the importance of not underestimating the continued significance of national governments with regard to shaping policy trajectories, particularly in states where there are weak layers of regional and local government. It also demonstrates there while there may well be some movement in policy making away from hierarchical (top-down or trickledown) models towards more complex, multi-stakeholder negotiations over policy decisions that include both state and non-state actors (Peters and Pierre 2001), such movements are not necessarily significant, stable or enduring. In addition while there is a system of governance for waste management involving different tiers of government and different spheres of society it is not a consensual governance system of equal partners, nor is it a static system of governance. In this case nation state decisions to define the local as the decision making scale for waste management planning, set within a weak regional framework, were revoked when those local scales failed to adhere to nationally-sanctioned and private sector defined solutions to waste management that included the development of incineration facilities for municipal solid waste. The national government sought to claw back control over decision making about waste management, without explicitly relocating responsibility for making unpopular decisions about waste facilities to the national scale, despite making clear public statements about waste facilities being in the national interest. This clawing back was a direct result of successful resistance movements who strategically scaled their opposition movements to generate an international space of engagement from which to mitigate their problematic relationship to the space of dependence, the site of the incinerator. While the structural power of the state to reframe legislative competencies has subsequently progressed the development of incinerators there remain alternative scalar strategies for resistance open to opponents of incineration.

A reading of these political processes, which characterizes the resistance movements as only localized and reactive to a developmental state is therefore not reflective of the intricate and evolving patterns of power, politics and intrascalar relationships that constitute conflicts over incineration in Ireland. It is in this sense that a more Foucauldian reading of waste governance is useful, a reading that places conflict and power at the centre of understanding political processes. In essence, as with the analysis presented here, there needs to be more attention to the range of things that affect waste governance, what Darier (1999, 15) has referred to as the 'vast and heterogeneous webs of social practices criss-crossed by relations of power'. Following Foucault's analytics of government, that means 'attending to the practices of government' and the 'conditions of governing' (Dean 1999, 28) when an activity, such as the introduction of municipal waste incineration, is called into question. Attention to the practices of waste governance also requires careful consideration of the spatial strategies that are inherent in those webs of social practices, particularly in this case amongst state actors seeking to impose selected waste management strategies and amongst civil society actors seeking to resist those strategies and assert the validity of alternatives.

This chapter has demonstrated that the management of waste in Ireland has been undergoing significant upheaval in the last decade such that it is now shaped by the interaction of various tiers of government and different spheres of governance. Such interaction of tiers and spheres has been identified in other contexts as 'multilevel governance', however while systems of multilevel governance have been proposed, particularly within the EU, as a means to establish consensus-based politics, waste management in Ireland remains infused with conflicts.

Chapter 6

Garbage Governance in New Zealand: Clean and Green?

Introduction

During the 1990s management of waste was becoming increasingly problematic for policy makers in New Zealand. Within the municipal waste stream volumes of materials generated at the household level continued to rise, but identifying sites for landfills to deal with that waste was creating tensions within communities and between different sectors of society. At the same time political changes and new frameworks of environmental legislation were beginning to impact on traditional local authority decision making practices that subsequently affected waste management. This chapter provides an examination of municipal waste management set within the wider socio-political context of environmental governance in New Zealand described in Chapter 4. It begins with an examination of the waste policy interventions that have emerged in recent decades. This is followed by attention to the interactions between actors and agencies from the spheres and tiers of waste governance. Consideration of the outcomes of the intersection between interventions and interactions are presented in the third section. Finally particular consideration is given to the role and use of the Resource Management Act (RMA) in debates about and contestations over the landfilling of waste.

Policy Interventions

Changes in environmental planning and local government have had a major impact on the way that issues surrounding household waste are constructed and managed. In particular waste collection and disposal is no longer solely the public sector enterprise it once was. In many locations the privatization of local government services, including waste collection, transfer and disposal, means that the private sector now controls a significant proportion of the waste stream. Indeed the Ministry for the Environment have reported that less than 10 per cent of municipal waste collection is completed by the public sector, with the vast majority of waste (around 80 per cent) being collected by private contractors who are contracted by local authorities (Forfás 2006). There are also small and medium sized community-based organizations operating on a not-for-profit basis that are becoming increasingly visible players in the field of waste management and resource stewardship influencing both modes of waste management and policy discourses (White and du Preez 2005). These changes have occurred alongside a radical transformation in the legislative frameworks for managing the environment and waste.

In a similar pattern to other developed countries waste management in New Zealand has progressed over the past two decades from a local system of waste disposal in town dumps to a more sophisticated and regulated procedure that seeks to develop recycling and the prevention of waste. Pressures to improve waste management practices have come from increased concern about impacts of waste on human health and environments as well as weak progress towards sustainable development (Seadon and Stone 2003).

In the past the main pieces of legislation affecting waste management were the 1956 Health Act and the 1974 Local Government Act, both of which focused primarily on waste collection and disposal from a public health and service perspective. By the late 1980s it was recognized that such a perspective ignored the growing production of waste and in 1990 the government set a target for 20 per cent reduction in solid waste production from 1988 levels by 1993. This led to the development of recycling programmes in some local authorities and national guidelines for the monitoring and management of landfill waste. The growing attention to waste was consolidated following the adoption of the RMA and its requirement to control the discharge of contaminants (including waste) into or onto land, air or water. The specific impact of the RMA for waste was the requirement to take account of the transferral of wastes from one media to another. So, for example, solid waste could only be disposed of to landfill after consideration of the effects that might have on air quality. However by 1992 the government had dropped its recycling targets and the Ministry for the Environment (MfE) was directed to negotiate waste reduction targets through voluntary agreements with business sectors.

In 1996 an amendment of the 1974 Local Government Act made it a legal requirement for territorial authorities to prepare waste management plans in order to promote effective and efficient waste management in their district, taking into account the economic and environmental costs and benefits of the proposed plans. The waste management hierarchy was proposed as the guiding framework to aid local governments with their prioritization of waste management options, but little detail was given on the required content of those plans. There was no duty on local governments to set goals and timelines for waste improvement actions and waste production per capita continued to increase (Boyle 2000). Within this loose arrangement, however, the Amendment did allow territorial authorities to allocate incentives or disincentives to recoup costs incurred in implementing the plan and it allowed for the construction of by-laws by local authorities to regulate the disposal, collection and transport of waste.

Although it was established that New Zealand's waste problem was large and growing the MfE acknowledged that it could not present a detailed picture of the extent of the problem because of inaccurate, inconsistent and incomplete data. In response a national waste database was set up to generate baseline information about the nature of waste within New Zealand in order to better plan for its management. It concluded that in 1995 only eight and a half per cent of solid waste was recycled even though 80 per cent of New Zealand residents had access to a recycling programme (MfE 1997a). It also established that there was no incineration of residential solid waste although some furnaces were used for burning limited amounts of medical and toxic waste and some waste biomass. Published at the same time, the 1997 State of the Environment Report also concluded that 'while waste management responses increasingly include recycling, cleaner production systems and higher landfill fees, total waste has increased, our landfill management practices are generally poor, as are our practices and attitudes towards managing hazardous waste' (MfE 1997b, 16).

The key problem areas for waste management in the country were identified and examined in a survey of waste management actors conducted by the Waste Management and Pollution Prevention Branch of the MfE in collaboration with Auckland Regional Council and Zero Waste New Zealand (Boyle 2000). Representatives from the public and private sector were involved in the survey as well as Māori tribes (Iwi) and NGOs. Overall the survey revealed concern about a lack of co-ordination and consistency amongst local waste management strategies. It was also felt that a lack of policy at the national level meant that little pressure was applied to industry to address waste issues unless resource consent was required under the RMA. Even when resource consents were required the judgements about conditions and permitted developments still, initially at least, lay with the local tier of government allowing for variable treatment across the country. In addition it was argued that the transference of responsibility for waste management to the local level was not matched by devolution of financial resources to undertake the waste planning remit or to employ suitably qualified personnel to make judgements about the impacts of waste. It was established that many waste actors wanted further direction, funding and detail with respect to the waste management programme in order to reduce waste production and improve the effectiveness of managing waste. Indeed Boyle concluded that

> national government has abandoned waste management and pollution prevention to the local authorities which are struggling to provide direction and undertake waste minimisation with little support or funding (2000, 525).

Therefore while the RMA considers the effects of waste management practices sub-national local authorities are increasingly looking towards community planning practices enshrined in the 2002 Local Government Act as mechanisms to deal with wider issues of waste minimization and prevention practices. Under the Act waste management plans were still to consider each disposal option, from most desirable (reduction) to least desirable (residual disposal), but there was also a requirement to include details on the promotion of waste minimization education; the provision of waste disposal facilities; the collection and transportation of waste and the allocation of costs related to waste management. The new act did however make it a requirement that all territorial local authorities must adopt a waste management plan, following a prescribed consultative procedure, by 30 June 2005. There was also scope for territorial local authorities and regional councils to share the planning, physical resources and facilities for waste management and an emphasis on the economic, environmental and cultural well-being of people within their territory for local government was articulated. Simple collection and disposal of waste for the lowest economic cost was no longer prioritized within central government guidance. The territorial local authorities were allowed to create by-laws for waste management for a number of processes including the regulation of waste collection, transportation and deposition and the collection of charges for the use of waste facilities. Unlike the other powers, there is no requirement that by-laws be exercised consistently with a waste management plan. As by-laws are legally binding in New Zealand they are potentially a powerful tool and hold sway over a management plan. However the use of such a mechanism in the waste field (specifically to introduce a local levy of waste) has proved highly controversial and legally problematic process.

There is at present no statutory role for regional councils in waste management under local government legislation. However the responsibility of regional councils under the RMA for the control of discharges of contaminants and hazardous waste, for the protection of coastal and water environments and in relation to their general obligation to achieve integrated management of the natural and physical resources of the region means that issues relating to waste management necessarily come within the remit of regional councils. Central government guidance supports co-operation and co-ordination between regional and territorial councils and between territorial councils within a region, although there is no formal requirement for authorities to work together in this way. However the most significant development, alongside these amendments to the Local Government Act, was the publication of the *New Zealand Waste Strategy: Towards Zero-waste and a Sustainable New Zealand* (MfE 2002). The Strategy built on the work of, and was influenced considerably by, the waste minimization and management working group's discussion document, *Towards a National Waste Minimisation Strategy* (MfE 2000). It is a non-statutory policy statement setting direction for waste reduction and improved waste management in line with the nation's sustainable development objectives. Published at the same time as the Local Government Act Amendment 2002 it was the product of wide-scale discussion and consultation, which included civil servants, politicians, consultants and key figures involved in zero waste activities in New Zealand.[1] The influence of these figures can be seen in the adoption of the phrase zero waste in the title of the waste strategy and also in the overall tone of the document. The overarching goal of the Strategy is stated as 'lowering the social costs and risks of waste, reducing the damage to the environment from waste generation and disposal and increasing the economic benefit by more efficient use of materials' (MfE 2002, 3) and it calls for effective legislation to encourage waste minimization and management. The Strategy also recognized that waste policies have tended to focus on end of pipe solutions for the disposal of waste rather than prevention, which has allowed waste production to continue to rise with economic growth. Despite this there was still no suggestion that consumption should be curtailed with the Strategy stating that 'we [as a society] don't have to avoid the products and services we normally use' (MfE 2002, 19), simply that New Zealand needs to 'produce more with less' (MfE 2002, 3). Such statements bring the ethos of the Strategy closely in line with practical applications of ecological modernization that have been identified in New Zealand's environmental policy by a number of commentators (Carolan 2004; Jackson and Dickson 2004). Most significantly the Strategy incorporates targets (see Table 6.1) for a whole range of wastes and five core policies including a sound legislative basis for waste minimization and management,

---

1   The Trust was established in 1997 with the aim of encouraging all sectors of society to reduce waste with the ultimate goal of achieving zero waste (Envision 2003).

efficient pricing, high environmental standards, adequate and accessible information and efficient use of materials. Table 6.2 details the key principles of the strategy.

Prior to the publication of the Strategy the MfE had predominantly relied on voluntary agreements and partnerships funded through the Sustainable Management Fund (SMF) to encourage waste minimization and resource recovery. While there had been discussion of a waste tax in New Zealand following the publication of the Strategy a decision was made to move away from eco-taxation and concentrate on other mechanisms for minimization. An emphasis on volunteerism remained the key approach with a commitment to funding voluntary community, industry and local government programmes and technical assistance for the waste industry. The review concluded that stronger leadership was required by central government and commitment was made to produce more extensive guidelines, establish standards and draft a national product stewardship policy. It was also decided that a crucial component of improving waste management was to provide information, and good channels of communication to disseminate that information, for all actors involved in managing waste. Improvements in waste data had begun to develop following the Waste Analysis Protocol established in 1992 and the Ministry for the Environment published a National Waste Data Report in 1997, but as mentioned earlier the scarcity and inaccuracy of waste data limited the usefulness of these initiatives. A survey of landfill management practices was conducted in 2002 and, while revealing significant improvements, it found that comprehensive waste data analysis had yet to be institutionalized on a national level. More successful, at least from an implementation perspective were initiatives to enhance community understanding of waste generation and management issues.

In 2003 the national government launched its first nationwide waste awareness campaign, a pilot project called *Reduce Your Rubbish*, which adopted a social marketing approach in an attempt to try and improve household attitudes and actions towards waste (see McKenzie-Mohr and Smith 1999). The campaign involved public information broadcasts using well known celebrities and humorous competitions to create publicity for waste reduction issues on national television; a campaign web-site and a hotline for households seeking advice and information from their local councils. The national event was co-ordinated with advertising and media coverage at regional and local levels in many locations. While clearly focused on waste matters the MfE were keen to emphasize that the project was also a pilot programme to 'test whether central and local government could work together to cost-effectively raise awareness of the waste issue and encourage householders to take some simple actions to reduce their rubbish' (Bradshaw 2004, no page number). The aim was to establish a partnership with regional government and the campaign drew heavily on the experiences of Auckland Regional Council's *Big Clean Up* exercise (Frame 2004). Research was conducted before the campaign started in order to set benchmarks for the post-campaign evaluation (Gardner 2003). The results indicated that the reach of the initiative was around 500,000 households with 20 per cent of survey respondents claiming that the campaign had a positive effects on their awareness, attitudes and behaviour (MfE 2005). The MfE and participating regional councils provided funding for the project, around $800,000 NZ, around half of which was spent on the national media events. Nonetheless these costs are relatively low in comparison to national waste awareness campaigns that have taken place elsewhere.

Table 6.1 New Zealand household waste targets 2002 (with progress review in 2004)

Target Area	Target Description
Waste Disposal	• By December 2003, local authorities will have addressed their funding policy to ensure that full cost recovery can be achieved for all waste treatment and disposal processes.
	Review: The Ministry for the Environment has developed a guide that can assist councils to calculate the full costs of landfills. The cost of collection needs to be incorporated into this model to enable councils to comply with this target.
	• By December 2005, operators of all landfills, cleanfills and wastewater treatment plants will have calculated user charges based on the full costs of providing and operating the facilities and established a programme to phase these charges in over a timeframe acceptable to the local community.
	Review: There is anecdotal evidence that many landfill operators are using the Landfill Full Cost Accounting Guide for New Zealand (Ministry for the Environment, 2002a) to achieve this target. However, no formal survey of uptake of the guideline or calculation of user charges has been undertaken. Based on this limited evidence it is estimated that most, if not all, landfill operators are on track to introduce user charges over a time frame acceptable to the local community. There are some significant remaining issues around charging at rural transfer stations and for domestic rubbish collections.
	• By December 2010, all substandard landfills will be upgraded or closed.
	Review: The results of the 2002 Landfill Review and Audit indicate that significant progress is being made towards this target, with the number of landfills in New Zealand decreasing and a trend towards higher standards in siting, design and operation.
Waste minimization	• Local authorities will report their progress on waste minimization and management for their annual report in 2000/02 and quantitatively on an annual basis from then onwards.
	Review: Most councils mentioned something about waste issues in their annual reports although reporting styles varied from place to place.

	• Ninety-five percent of the population will have access to community recycling facilities by December 2005. *Review: Ministry information suggests 90% of the population had access in 2003 and all councils were on target to meet the 2005 deadline.* • By December 2005, territorial local authorities will ensure that building regulations incorporate reference to space allocation for appropriate recycling facilities in multi-unit residential and commercial buildings. *Review: A small number of councils now stipulate this in their local plans, but this was recognised as an inefficient way to meet the target and discussion is being had to incorporate this requirement into the Building Act and building regulations.*
Organic	• By December 2003, all territorial local authorities will have instituted a measurement programme to identify existing organic waste quantities, and set local targets for diversion from disposal. *Review: Pilot work has been undertaken in a few councils but there are issues about commercial sensitivity of data on private collections of organic material and concern has been expressed about the lack of commercial incentive to compost organic waste.* • By December 2005, 60 percent of garden wastes will be diverted from landfill and beneficially used, and by December 2010, the diversion of garden wastes from landfill to beneficial use will have exceeded 95 percent. *Review: Many councils provide the opportunity at their landfills or transfer stations for the diversion of garden wastes for composting or mulching. However, green waste going to landfills still provides a significant fraction of total waste (up to 25 percent in some areas) and there are remaining problems about the existence of beneficial end uses for the products generated from garden wastes and contamination of compost is a significant issue, both in New Zealand and internationally.* • By December 2007, a clear quantitative understanding of other organic waste streams (such as kitchen wastes) will have been achieved through the measurement programme established by December 2003. *Review: Similar issues to target 2.1. Achieving the target requires work to quantify the different elements of organic wastes.*

Source: Ministry for the Environment (2002; 2004).

Table 6.2 Key principles of the 2002 New Zealand waste management strategy

Principle	Description
Global citizenship	• The effects of waste are not confined to localities • New Zealand must take responsibility for the global consequences of waste
Kaitiakitanga (Stewardship)	• Everyone is responsible for looking after the environment • Māori believe all living things are related and that kaitiaki (stewards), are obliged to maintain the life-sustaining capacity of the environment for present and future generations
Extended producer responsibility	• Those who make goods and deliver services should bear some responsibility for them and any waste they produce, throughout a product's entire life-cycle
Full cost pricing	• The environmental effects of making, distributing, using and disposing of goods and services must be properly costed and charged where they occur
Life-cycle principle	• Goods should be designed, made and managed so all their environmental effects are accounted for and minimised, until the end of their lives
Precautionary principle	• A lack of scientific certainty must not be used as a reason for ignoring serious environmental risk

Source: *New Zealand Waste Strategy* (2002).

Local Government New Zealand, who acted as a broker between national and regional governments in the pilot project, surveyed the participating councils after the campaign and established that overall participants were generally positive about the partnership arrangement. The reflective approach adopted in the campaign culminated in a number of recommendations which acknowledged that while providing good immediate results such intermittent one-off campaigns have limited impacts in the long-term and that awareness campaigns require sustained initiatives articulated within localities to derive continued impacts on attitudes and behaviour. In response more locally-focused initiatives, such as the Sustainable Households Programme, are being developed by the not-for-profit sector often in collaboration with territorial authorities (Taylor 2005).

Reflecting on the progress made in waste management, *Waste Minimization in New Zealand: A Decade of Progress* (MfE 2005) identifies the marked improvement in landfill standards and the provision of kerbside recycling schemes. It also accepts however that results have been mixed in terms of councils introducing cleaner production programmes, producing high quality waste management plans and in terms of activating voluntary agreements with industry. It seems that the difficulties New Zealand faces in relation to its waste management challenges are both general and specific to its socio-geographical context. The more familiar barriers relate to inaccurate pricing, unreliable markets for recyclable materials, limited information on the extent of the waste management problem and a poorly defined understanding of production and consumption trends that influence the amount of waste generated. However New Zealand also has particular characteristics that create unique waste management challenges such as its isolation from the rest of the world and its reliance on imports for many consumer products. This works in conjunction with the dispersed population to increase the costs of transporting waste and recyclates. In response to these challenges a Waste Minimization (Solids) Bill has been proposed by a Green Party MP to, amongst other things, establish a centralized agency for facilitate progress in the waste arena. This Bill is currently being considered by a government Select Committee which is due to report in 2007.

The evolution of waste management policy initiatives described in this section indicates the nature of the governing rationalities in New Zealand, the dominant mechanisms that have been adopted, the technologies of waste governance, and the primary agencies involved in governing waste. A summary of these interventions is listed in Table 6.3. However the question as to how those modes of governance emerged remains. In order to get a deeper understanding of waste governance systems it is necessary to look behind the statements and legislation to the interactions between spheres of governance operating at a range of scales.

Interactions

New Zealand's geographical isolation and political autonomy, combined with the lack of strong global waste governance structures, means that there is little overt supra-national pressure affecting its waste management practices. Nonetheless New Zealand is a signatory to international initiatives, such as the transport of hazardous waste through the 1989 Basel Convention on the Control of Transboundary Movements of Hazardous Wastes and their Disposal (ratified by New Zealand in 1994) and the 1986 Convention for the Protection of the Natural Resources and Environment of the South Pacific (ratified in 1990). The government also maintains an eye on the policy initiatives emanating from the EU and Australia recognising that

> New Zealand's clean green image is useful in promoting tourism and exports. Poor waste management would damage both the reality and the image. Our record on waste has important implications for trade and tourism, and sustainability of all New Zealand businesses (MfE 2002, 17).

Table 6.3 New Zealand waste policy interventions

	Title	Date	Key Details
Policy Documents	New Zealand Waste Strategy	2002	• First comprehensive plan covering solid, liquid and gaseous wastes, and dealing with waste from generation to disposal including: promotion of materials and resource efficiency; provisional targets and standards; information and communication; full cost accounting of waste facilities
	Review of targets in the New Zealand Waste Strategy	2004	• Establishes usefulness of targets and argues for no change to existing targets • Claims good progress has been made towards some targets • Identifies problematic areas of little progress towards targets e.g. organic waste and information deficit problems with regards to private control of waste stream • Establishes that an effective and cost-efficient monitoring and reporting system is essential for measuring progress
	Waste Management in New Zealand: A decade of progress	2005	• Details increase in waste legislation • Identifies increases in waste plans and recycling services at the local and regional level • Reports on improvements in landfill operations

Policy Initiatives	National Waste Awareness Programme (Reduce your rubbish)	2003	• Joint central and local government pilot campaign
			   o Business and household focus
			   o Mass media campaign
			   o Website information

	Govt[3]	2003	• A voluntary programme which aims to help central government agencies improve the environmental sustainability of their activities including waste management

Category	Instrument	Year	Description
Legislation	Resource Management Act	1991	• To ensure environmental impacts of development are managed to reduce the effects on the environment, including waste management facilities. In 2004 14 national environmental standards were developed under the Resource Management Act. Those relevant to waste include: banning activities that discharge significant quantities of dioxins and other toxics into the air; ambient (outdoor) air quality standards; and a requirement for landfills over 1 million tonnes of capacity to collect and destroy greenhouse gas emissions
	Local Government Act (Amendment of 1974 Act)	1996	• Direction to territorial authorities to produce plans • Allows for development of bylaws for the collection, disposal and management of waste • Allows for incentives/disincentives to recoup costs of waste management plan
	Local Government Act (Amendment of 1996 Act)	2002	• Including a requirement for territorial authorities to complete waste management plans by 30 June 2005
	Waste Minimisation Solids Bill	2006	• Referred to the Local Government and Environment Select Committee in Parliament in 2006 • Calls for: establishment of Waste minimisation Authority and Waste Control Authorities; a levy on all waste sent for disposal; public authorities to implement green procurement policies; extended producer responsibility schemes for certain products; ban on certain materials going to landfill
Policy Instruments	Producer Responsibility Schemes	From 1996	• Voluntary agreements including: Packaging Accord; WEEE; tyres; waste Oil; paint; refrigerants
	National Waste Database	1997	• First and only national waste database published
Funding Schemes	Sustainable Management Fund	1994	• Funds projects that strengthen proactive partnerships between the community, industry, iwi and local government • Between 1994 and 2005, funding for waste-related projects equated to around 36% of the total fund allocation over that period

Despite this awareness of, and engagement with, international activities waste policy has tended to be driven forward by internal conditions rather than exogenous pressures. These pressures emanate from a variety of sources across public, private and civil society sectors and vary according to local contexts and conditions. When the New Zealand waste strategy was launched central government claimed a commitment to sound waste minimization and management as a key element of its movement towards sustainable development and identified positive inter-scalar relations between central and local governments in this regard (MfE 2002). However the quality of relationships between central and local government have been questioned and the hands-off style of central government that has come to dominate policy landscapes since the neo-liberal reforms of the 1980s has led to significant variations in local authority practices across the country.

While national and sub-national governments are undoubtedly important in framing and implementing waste legislation it is the private sector that dominates the collection, transport and disposal of waste in New Zealand. As detailed earlier the scale of private operations, including key players Waste Management New Zealand Ltd., EnviroWaste Services Ltd, Onyx and Perry Waste, means that control of the waste stream from door to dump is predominantly privatized. However while the private sector is unquestionably dominant in terms of the volume of waste managed it is the community sector that provides the largest number of waste operators across New Zealand and the influence of this sphere of governance is reflected in the adoption of zero waste rhetoric within the national strategy. The picture of waste management in New Zealand then is not one of simple neo-liberalism with a hollowed out central government, private sector service delivery and a lack of regulation; although these features are present. The spatial signature of waste management is actually much more complex with a variety of individuals, organizations and governance spheres seeking to exert some influence on how waste is perceived and managed. As detailed in the previous chapter, when considering the waste governance of Ireland, it is clear that while influence can be inferred in terms of the wording or direction of policy statements and legislation these rarely tell the full story of participation or exclusion, negotiation or resistance that lead up to these positions, or indeed affect the implementation (or not) of policy intentions. Insights into these processes can be gleaned in part from media coverage and parliamentary debates and these were considered, while additional information based on interviews with actors from across the spectrum of waste governance also provide the backbone for the analysis here.[2]

When resource management reforms began in the mid 1980s there was broad agreement between business and environmental interests that systems of management were constrained by the plethora of uncoordinated and fragmented legislation. Both called for more decentralization and accountability in governance approaches but there were significant differences in the reasoning behind these similar positions (Bührs and Bartlett 1993). Business sought a smaller, less regulatory government that supported privatization of resources and allowed resource development to

---

2   Thirty interviews were held with local, regional and national government officials, with private sector waste consultants, academics, waste companies and representatives of not for profit waste organizations.

occur through the market (Harris and Twiname 1998). Environmentalists wanted more controls to facilitate environmental protection, a separation of development and management responsibilities in government agencies through more devolved participatory democracy and were opposed to privatization of public resources (Ericksen et al. 2004). While both sets of lobby groups and political parties envisaged a greater role for local authorities there was disagreement about which model of reform, decentralization or devolution, to follow. Eventually the RMA specified reserved powers for national government to establish policy statements and regulations on issues of national importance, while sub-national governments were required to produce plans including regulatory rules. A three-tiered system for administering environmental planning, including waste management, involving central, regional and local government was then established. While national government could get involved in local affairs this was expected to be only a rare occurrence and sub-national councils were free to adopt 'as much or little regulation as the local circumstances would allow' (Ericksen et al. 2004, 24). The system was to operate through a partly devolved system that emphasized flexibility, partnership and capacity building to develop plans for sustainable development while at the same time adhering to the principles of free-market liberalism.

Yet, as detailed in the previous section, there was little in the RMA to shape the trends in waste production. The effects-based emphasis of the Act focused attention on the impact of the facilities developed to deal with waste rather than mechanisms to minimize its production. In the absence of strong guidance from central government on waste minimization, applications for resource consent from waste companies became a focus of attention for environmentalists and local communities concerned with the trajectory of waste management practices. Effectively the implementation of the RMA became a site of contention and battles over waste were, for the most part, played out in localities based on legal and scientific criteria regarding waste disposal. During the late 1990s pressure from environmentalists, and particularly advocates of zero waste approaches, sought to rescale and reshape these arguments with calls for stronger national attention to waste minimization. The result of these debates led to the formation of a waste strategy *The New Zealand Waste Strategy: Towards Zero Waste and a Sustainable New Zealand* in 2002. The adoption of zero waste discourses, at least in the title of the document, was a significant coup for supporters of the zero waste philosophy and was identified as a threshold moment for waste management. However as the following analysis indicates the rhetoric of zero waste is not universally supported. First attention is given to the perspectives of actors from all spheres of governance regarding waste governance practices. This is followed by consideration of inter-actor relationships in a specific case study of waste management debates that revolve around the development of a landfill under the RMA.

Inter-governmental Relations

When the New Zealand waste strategy was launched central government claimed a commitment to sound waste minimization and management as a pivotal element of its movement towards sustainable development and identified positive inter-scalar relations between central and local governments as key to achieving this goal (MfE 2002). However the non-interventionist style of central government that has dominated policy landscapes since the neo-liberal reforms of the 1980s and the introduction of the RMA have led to significant variations in local authority practices across the country. Some have viewed this variation positively in its permissiveness of locally generated actions.

> We are just lucky I suppose that we still have a local authority structure with empowerment to do something (Private sector 5).

However the waste actors interviewed more commonly expressed concern that variations are shaped more by access to resources than by any pro-active sense of subsidiarity. These concerns were expressed by elements of the private sector, by community organizations and by local authorities themselves. The decentralization to the sub-national level of decisions over waste facilities through the RMA and the flexible interpretation of the RMA legislation was particularly singled out for criticism. As one regional officer commented

> in well managed and well resourced authorities it works well, some authorities do struggle and so the application [of the RMA] is patchy. Without a significant rating base it's difficult to employ enough people with enough skills and experience to cover all that is required (Regional public sector 3).

A landfill manager from the private sector reiterated these issues by saying

> there are clearly some defects with it [the RMA], one is that it is not centrally run by the government, it is run by different regions and the interpretations are vastly different (Private sector 2).

The difficulties of variable standards were seen to permeate through the RMA process. While several commentators, like those above, mentioned the uneven practices in granting consents others referred to the irregular approaches to monitoring and enforcement.

> I think in one place someone visited our site once in ten years. They just don't go out on site as long as you don't get complaints ... in other places it's very different. Enforcement is regional, depending on where you are in the country, the regional council might be a couple of young graduates and their manager might be a person who has been waiting for their pension for the last ten years. He doesn't want to rock the boat, because all his buddies are out there on the farms and he wants to keep his job (Private sector 1).

A lack of leadership from the Ministry for the Environment was seen as the primary cause of the unevenness in RMA action.

> They [the MfE] are really toothless, for example on landfills, they came out with a statement that all sub-standard landfills should have a resource consent within two years of the introduction of the RMA, that's back in 1994, there are still landfills today operating without consents (Private sector 3).

Comments from individuals across all spheres of governance suggested that greater direction from central government would address the variable interpretation and implementation of the RMA. The private sector commentators wanted national enforcement practices and standards to ensure common treatment of waste facilities in different locations and most envisaged a single enforcement agency, similar to the EPA in Ireland and the USA and the Environment Agency in the UK, would be the most efficient means of achieving this goal.

> There's a lot of people in the private waste industry that truly believe there should be an Environmental Protection Agency system in New Zealand ... we don't have a level playing field (Private sector 2).

Others felt, however, that the possibility of establishing such an agency in the current political climate of New Zealand was highly unlikely.

> We have tried for years to get the government to have a central directive, but New Zealand has had a hands-off style of government for years on everything (Waste network 2).

While the lack of leadership at the centre was identified as a weakness in the devolved RMA system because of the potential for variations in interpretation and implementation, the situation was perceived as even more problematic in relation to waste minimization where it was felt that the absence of legislation undermined the initiation of, and support for, such activities. Sub-national tiers of government voiced these concerns most frequently.

> The MfE, I think, have just tagged along. That's all they've done. They keep producing this guff, but at the end of the day I believe they have done very little to promote waste minimisation in New Zealand ... there is no legislation that they have managed to put in place that has assisted us (Local public sector 3).

However both community groups and the private sector pointed to the lack of progress towards zero waste, as enshrined within the waste strategy, as a key example of a vacuum at the centre of government.

> The national waste strategy has all the good words in it, but they are so averse to anything that represents an intrusion of any sort, whether it's a tax or a levy, that affects big business. It's become so free market (Community sector 5).

The reluctance to develop a national landfill levy was also seen as an indicator of weak national commitment to mandatory techniques aimed at minimising waste.

> [The MfE] should bring in brave things, put a landfill levy on, use it to fund the things you want to achieve. At the moment the landfill companies have responded to the growing calls for landfills to be properly costed by increasing the price of landfilling. They're reluctantly saying they want a national landfill levy, but they are trying to prevent local ones being put on. They know the government won't do a national one so they can clamour for that while the local councils are trying to put them on because they're sick of waiting for the national government to do it (Private sector consultant 1).

Officers in the Ministry admitted that their ability to take a strong lead in these issues was constrained by their primary role as guides rather than dictators in waste management planning process, but equally there seemed to be no strong desire to change the status quo despite pressures from civil society and the private sector. Public sector officers were candid in their assessment of the reasons behind the approach of central government. On the one hand there were allusions made to the persistence of a free market approach.

> I just don't think they want to be seen as draconian ... They don't want to upset the money maker ... They want to do things voluntarily (Local public sector 2).

On the other they identified practical constraints.

> We are asking for legislation, but there's no legislative drive in MfE. It's a pro-industry government. We need firm national guidance in terms of targets and monitoring, but they've no teeth, no resourcing, no staff time and no accountability (Local public sector 1).

Rather than strengthen the centre, national government is encouraging greater regional co-operation between councils for waste planning as a mechanism to try and bring laggard local authorities up to higher standards and create more cohesion in waste practices. Yet again however these moves have not gained widespread support, at least from the local authorities interviewed.

> There's a strategy for this region. You wouldn't pick it up and find anything useful to do in it. That's a drawback of our system. Other places can make a decision overall, whether it's good or bad whereas here there's no overall policy to speak off and as a result we're all at different stages (Local public sector 2).

The concern is that such regionalization without strong legislative backing, standards or targets will not have the legislative weight to improve practices as recognized by the Ministry,

> we are encouraging joint strategies, regional plans, but these are non-statutory documents and no consultation is necessary. It is effectively a linking document ... the plans don't go down to a level of details for infrastructure provision, they're more a framework for dealing with waste (National public sector 1).

Such an approach presupposes good relations between regional and territorial local authorities, adequate skills and resources and a strong commitment to waste matters. Where these exist positive outcomes, innovation in waste management at the local level and significant diversion from landfill, have been reported.

> A lot of people ask us, how come we're doing so well with recycling. It's because we forced it to happen ... and we weren't the most popular people here at the time (Local public sector 3).

Yet there was a strong feeling that as the waste industry becomes increasingly privatised, even local authorities with commitment and resources may find it increasingly difficult to affect waste policy outcomes.

What we've seen is that unless you can control the waste stream from go to woe the council loses its way because it's being driven commercially. If you can't control the process you can't control the outcome (Local public sector 1).

Non-governmental Waste Actors

So while national and sub-national governments are undoubtedly significant in framing and implementing waste legislation it is the private sector that dominates the collection, transport and disposal of waste in New Zealand. Such is the dominance of some private waste companies that concerns about monopoly and anti-competitive behaviour through their acquisition of smaller companies around the country have been mooted (Commerce Commission 1999).

In interviews private sector actors saw themselves as de facto watchdogs given the absence of a central regulatory institution. Indeed private waste industries had participated in RMA consent processes where they had no direct interest in order to try and ensure that similar consenting standards were maintained. One actor, who had previously worked for the Ministry, but who was now employed in the private sector, reported that on occasion government officials would contact large waste companies and let them know that a sub-standard application had been lodged stating that they would support the waste company if they decided to make a submission, but that the ministry would not be submitting because of costs.

> So we would go in and submit against it in order to try and make a level playing field ... it was a matter of principle, we are not going to let anyone else have poor quality facilities because look what we've had to do. Our actions were seen as anti-competitive, but in effect it wasn't ... we have a very strong conscience when it comes to environmental issues (Private sector 2).

While in this case business interests of the large private sector companies were served by calling for enforcement of common conditions for RMA consents there was a general lack of enthusiasm for more regulation and support for voluntary mechanisms amongst the private waste industry in other areas of waste minimization. An argument against more regulation was the perception of waste minimization as driven more by politics than by science.

> The biggest challenge is the lack of science and the lack of fact ... I get so frustrated that decisions are made and there's no substantial or robust or rigorous scientific backing to it. That's where most of us come from, the science background and decisions shouldn't be made unless there is robust and rigorous information and analysis behind it. A lot of Wellington's [MfE] policy decisions, across the board in different departments, are based on who you know, who you have coffee with, the personalities and it's not good enough (Private sector 3).

Interviewees from the community waste sector were most concerned about the dominance of a profit motive over waste minimization in the increasingly privatized waste arena, particularly amongst the larger private players. These commentators suggested that it was mass cartage that drove private sector operations and, in the absence of landfill taxes and in the presence of local authorities with financial constraints, landfill disposal would remain the preferred option. Not for profit waste actors complained that in

> Privatising waste we have locked waste away from the community. You've got to be able to access the waste stream, the big companies will use it as a profit making operation, however the community would look at it as providing a wider range of services (Waste network 1).

More bluntly another actor claimed

> They're still at the truck it and dump it stage ... we need these companies to evolve into recycling companies (Waste network 2).

Fundamentally there was a concern that the consolidation of the waste industry was working against resource recovery.

Interestingly, the Forfás (2006) figures indicating private sector control of the waste stream actually include community based not-for-profit organizations in the privately-owned category and there is no detailed breakdown between commercial and community based operations. Such figures are not held either by central government or networks such as Zero Waste New Zealand. While it may be confidently assumed that commercial sector is dominant in terms of the volume of waste managed there is some evidence to suggest that the community sector actually provides a larger number of waste operators (albeit of smaller scale) across New Zealand and non-governmental initiatives are increasingly being recognized as important areas contributing to waste minimization (see Envision 2003b; Stone 2002; White and du Preez 2005). As Stone notes 'the growth in non-governmental organizations that directly or indirectly promote waste minimisation are related concepts has been phenomenal over the past decade' (2003, 18).

That the language of zero waste was adopted in the first New Zealand waste strategy was identified by the majority of actors within the civil society waste sector as a major victory. In particular it was the work of key zero waste advocates, who had been involved in setting up the Zero Waste New Zealand Trust, who had managed to get the ear of officers in the Ministry for the Environment during the late 1990s and early 2000s. The Zero Waste New Zealand Trust is a charitable trust established in 1997 through support from the Tindall Foundation. More recently a range of sources including the Sustainable Management Fund, a central government fund, have supported Zero Waste. The focus of the Trust is to support the activities of community organizations, councils, businesses, schools and individuals involved in waste minimization and recycling and it has achieved this through a variety of mechanisms from information exchanges through newsletters, website, conferences and reports to research, education, funding and advocacy. It has been successful in encouraging all (bar one) South Island authorities and a large number of North Island authorities to adopt Zero Waste policies.

That the discourse of zero waste was adopted in the Strategy was not welcomed by all waste actors however, particularly those in the private sector. One representative from a large waste company suggested that

The group who developed the strategy weren't elected people, they were generally the loudest people ... The Ministry was in a hurry, they picked a whole lot of people who they thought should be on the group and then those people came on and the people with the strongest voices got whatever they wanted put into the strategy ... the language was put in there to try and appease the zero waste people (Private sector 3).

Another private sector employee argued against the very notion of zero waste.

I prefer the term waste minimisation, because your goals have always got to be achievable and zero waste I think just puts out the wrong idea (Private sector 1).

Ironically the language of zero waste was seen by some as the reason why activity in the waste arena had not progressed at the same pace since the introduction of the Strategy. As one interviewee saw it

The strategy just ended up being a document that was rushed through, no one who wrote it really believed in it except maybe a couple of people and then Barry Carbon [the then newly appointed CEO of Ministry for the Environment] came on board and realized that it was a bit of a joke and cut the budget by half in the waste group (Private sector 2).

A change in leadership led to restructuring at the MfE. The waste unit was dissolved and attention shifted to permeating waste issues throughout the work of the MfE. While some welcomed this shift other felt that it in effect diluted pressure to change practices.

The flurry of activity between 2000-2002 has quietened down ... waste dropped down the agenda. There are targets in the Strategy, but they are of little use and little action has been taken. It's in a state of flux, monitoring of waste is still not in place, it's not funded by the MfE and local authorities are not funding it (Private Waste 5).

While the zero waste network is still in action and its members still pushing the agenda one key individual in the movement suggested that

it hasn't become a popular movement, like with people waving banners 'down with waste', but I think it's got into the psyche a bit ... I've got enough people to have changed the waste paradigm for ever ... look at the plans, they've got zero waste at the top as a goal. It might seem like tokenism to some degree, but that empowers people and local activists reinforce it (Community sector 6).

Yet, as seen recently in the changing leadership of the MfE and with it the changing emphasis, civil servants can also influence the way that waste is governed and while getting certain discourses in place is important it does not mean that action will necessarily follow. As a supporter of zero waste said

you look at the five year plan at the MfE and it's written with marshmallow, inside a duvet, there's nothing in there. It all sounds good, but in five years it will look exactly the same. I'm an optimist, but I have to say I'm disappointed with the way it is (Community sector 3).

Civil society organizations such as Zero Waste New Zealand (ZWNZ) and The Waste Exchange are networks that operate at a national level while others, such as Wastebusters Canterbury or Innovative Waste Kaikoura, are community-based resource organizations that have a more local focus. Some organizations do not specifically target waste issues although they come within their wider remit of environmental or sustainable development (such as the Sustainable Business Network or the Environmental Defense Society) and others emphasize particular tools, approaches and mechanisms to address waste related issues (e.g. Kaitaia Community Business and Environment Centre and The Natural Step Environment Foundation Aotearoa New Zealand). Stone (2003) suggests that the structure of NGOs contributing to waste minimization and related activities has evolved from issue specific campaigning pressure groups towards more collaborative organizations seeking to provide information and networking options for all actors willing to move towards more sustainable practices. This position is supported by actors in the sector such Zero Waste New Zealand who see that the community sector has tried to fill the vacuum resulting from the neo-liberal governing style of the 1980s by taking ownership of problems in their communities (Zero Waste New Zealand, 2003a).

Since the radical political reforms of the 1980s the number and variety of different NGO organizations actively pursuing waste minimization has grown considerably, but a particularly dynamic grouping within these NGOs are community-based recycling organizations who are directly involved in resource management for localities. Often these groups will benefit from public sector assistance through funding or subsidies for land or buildings (Stone 2002). The majority of these organizations in New Zealand focus on solid waste, particularly domestic solid waste and activities such as kerbside recycling services, management of reuse/recycling centres and shops, educational programmes for schools, publics and businesses, rehabilitation work experience for people with special needs or the long term unemployed. In some cases, such as Innovative Waste Kaikoura in the South Island, organizations may manage all waste services for local communities including collection and landfill management.

Research conducted by White and du Preez of 18 community-based organizations in New Zealand found that, although often being committed to more than one objective, most groups felt that environmental benefits were the prime motivation for their activities (78 per cent), followed by education, community involvement, social capacity building and finally campaigning (2005, 15). Importantly community-based organizations in New Zealand have tended to generate high levels of support amongst local people as communities acknowledge the wider social, economic and environmental returns they receive from these organizations compared to public or private sector waste management companies. It is acknowledged by waste actors across the spheres of governance that the community-based organizations have had more impact outside the big cities in New Zealand, in locations where there has not been competition from big companies and in areas where the councils were not active in waste minimization. It is also accepted that the groups have been pivotal in raising awareness of alternative waste management processes outside landfill disposal and demonstrating the importance of community participation in waste issues. Equally the organizations have provided additional employment, often for people who may have had difficulty getting work previously.

However, in contrast to locally supportive environments, the relationship between central government and community-based groups has not always been so positive (White and du Preez 2005). In particular the decision to cease the operation of the Community Employment Group was seen as undermining an important source of support for the groups both financial and in terms of knowledge about the needs of community-based organizations. It was felt that the MfE did not recognize the added benefits that community-based groups brought to local communities and it was feared that they saw the community sector as just another waste company. Although New Zealand's community sector is specifically mentioned in the Waste Strategy it is afforded only the rather ambiguous statement that

> the sector is expected to continue its work while looking for ways to do it better'. By working closely with other sectors, it can share resources to maximize effectiveness ... groups in this sector should think carefully about their purpose and focus on what they can deliver most effectively (MfE 2002, 44).

There is no guidance on how those relationships might be forged or how resources might be shared and there is no explicit commitment from the MfE that they will specifically provide resources to assist the community sector in their attempts to forge new relationships. In response there were increasing calls from civil society groups to create a national umbrella organization that could communicate the needs of this sector more effectively alongside private sector interests in central government policy making. To this end the Community Recycling Network was established as the representative organization of existing and emerging community groups involved in moving towards zero waste with the aim of working in collaboration with other organizations such as Zero Waste New Zealand and in partnership with local authorities.

In contrast to the overwhelming concern about central government's support and commitment to the community waste sector the relationship with local authorities appears to be more variable across groups. In White and du Preez's survey 50 per cent of the groups involved said they had positive relationships with their territorial local authority (2005, 20). While the practices of some local authorities were seen as a limiting factor for groups some commented on the positive, mutual learning processes that both community-based groups and local authorities were embarking upon. As with discussions about central government's relationship with the community sector it was felt that territorial authorities needed to recognize the added value of their operations above and beyond those currently provided by public and private sector waste service providers.

It is difficult to evaluate the quantitative successes of the community waste sector when there is often only limited information available on volumes of waste being diverted from landfill or amounts recycled. The measurement of success in terms of quantities of recyclable materials processed is more commonly used amongst community groups. There is also the difficulty of measuring more qualitative aspects of community based resource organizations such as enhancing community pride, generating local empowerment or changing the actions and culture of communities. As Stone notes (2002) it is often difficult to set appropriate and feasible targets in these areas.

Despite the considerable challenges that the community sector face in New Zealand the number of organizations continues to increase as does the percentage of the waste stream that such groups deal with. Although it can be assumed that the community sector's contribution is still small in relation to the volumes treated by private companies they are increasingly becoming important players within the New Zealand waste management landscape. Part of their significance has to be assigned to the passionate belief that community-based organizations have in the value of their work to society as well as the environment. Indeed the commitment to the goals of waste reduction and community empowerment of key figures leading community waste organizations has been recognized as an important ingredient for community groups. In New Zealand it has been demonstrated that these key figures, or charismatic champions, can generate a public and political profile for community operations.

In sum community waste projects all work to promote the vision of the New Zealand Waste Strategy and many are explicit about their commitment to reduce waste to landfill with the ultimate aim of Zero Waste. As a means to achieve this aim community groups have participated in the development of local waste management plans in localities and have been at the forefront of many innovative developments to improve resource stewardship within communities. Organizations such as Zero Waste New Zealand have been advocates of the community sector at a national level and have suggested that while 'community pioneers have been under-funded and in the past, often dismissed as fringe elements' (ZWNZ 2003a, 9) they are now in the ascendance.

The consideration of interactions between state and non-state actors in New Zealand reveals a devolved and decentralized system of waste governance. Following the neo-liberal reforms of the 1980s responsibilities for planning and management of waste have been directed downwards to sub-national levels of government and outwards to the private and civil society sectors. The absence of supra-national government and a weak lead from central government has facilitated the development of a variegated waste management and planning landscape. The private sector appears to be the dominant service provider and, through its active participation in the RMA, is a key player in decisions about waste management infrastructure developments. However civil society actors have, at times, also played a significant role in shaping discourses of waste in policy making. The emergence of the zero waste concept in the national waste strategy, which was fostered by the practices of the Zero Waste New Zealand network, being a key example of such civil society influence. In addition civil society organizations are providing opportunities for alternative waste management, or resource management, practices within specific geographical contexts across the country.

Waste management policy interventions are the product of tiers and spheres of governing actors and agencies interacting at particular moments and in particular spaces in order to influence their form and function. These processes are clearly important in shaping the outcomes of policy interventions and the following section examines the impact that the complex and shifting landscape of policies, programmes, rationalities, agencies and mechanisms have engendered.

Outcomes

If we are to give New Zealanders the environment they expect and deserve then a sound waste management programme has to be high on our list of priorities. Management of New Zealand's waste is not just an abstract environmental issue: it can affect where we live, work and play. Managing our waste in a sustainable and responsible way is a social and economic imperative. We owe it to ourselves and to future generations to get it right (Barry Carbon, CEO, Ministry for the Environment, in MfE 2005, 3).

Attention to waste governance has been on the agenda in New Zealand for the past decade. In part the flurry of activity on solid waste management at this time can be traced back to growing volumes of waste production and the difficulty of acquiring resource consent to construct new landfills under the RMA, as well as an increasingly vocal community and NGO sector proposing waste minimization approaches. It was a culmination of push and pull pressures that led to the development of a national waste management strategy which has been the guiding framework for reduction, recovery and better management of waste since 2002. But what impacts have the policies and programmes and the interactions of agencies and mechanisms had on the waste governance landscape? This final section considers the outcomes of New Zealand's waste governance and reveals divergent readings from different actors within the governing arena.

Good progress

As a precursor to the OECD environmental performance review that is due to report in 2007 the MfE conducted an appraisal of progress in the waste management arena (MfE 2005). This report presents a positive picture of developments detailing significant achievements across institutions and regulatory frameworks, waste reduction and materials efficiency activities, information and communication strategies as well as performance standards and guidelines. In particular the provision in the 2002 Local Government Act that all territorial authorities were to have waste management plans in place by June 2005 is cited as a major development in the field. In March 2005 more than 80 per cent of local authorities (82 councils) had waste management plans and 74 per cent of territorial authorities have included waste targets set out in the Strategy in their waste plans. In addition a set of national environmental standards have been introduced to the Resource Management Act and several of these have implications for household waste management activities, in particular the management of emissions from landfills and incinerators (should municipal solid waste incinerators be proposed in the future). There are seven standards banning activities that discharge significant quantities of dioxins and other toxics into the air and five standards for ambient (outdoor) air quality that set 'bottom lines' to protect people from harmful particulates. There is also a requirement for landfills over one million tonnes of capacity to collect and destroy greenhouse gas emissions.

Before the development of the *New Zealand Waste Strategy*, waste management planning was undertaken largely by individual councils and focused primarily on the role of councils in managing solid waste. However the process of developing the *New Zealand Waste Strategy* underlined that waste management was a trans-local matter and that its successful implementation would depend on coordinated action. The MfE have opted to support a regionalized approach to waste management, encouraging sub-national local authorities to work together in planning and regulating waste activities. Greater co-operation emerged particularly through the closure of local dumps and the development of large-scale, engineered regional landfills, but also increasingly through the development of regional waste strategies (e.g. Environment Waikato). Within this more co-operative governmental framework the MfE sees its role as a conduit for information for to local authorities rather than anything more prescriptive and to further this end a waste management planning facility on the MfE website has been developed to provide a medium for the exchange of good practice and debate between local authorities. Despite the acknowledgement of the need for some national standards there remains commitment in central government to the view that 'waste management planning at the local *and* regional levels provides the clearest basis for identifying both the local and regional waste management policies and priorities, and the actions needed to implement these' (MfE 2005, 16).

In addition to the development of greater trans-local governmental interaction there is evidence of greater co-operation between state and non-state actors in the waste management arena. In part this is due to the increasing levels of sub-contracting to the private sector in terms of collection and disposal of waste, but it has also resulted from the development of voluntary agreements and partnerships with industry. Examples include support for waste minimization activities such as the Packaging Accord and through funding for cleaner technology developments through the Sustainable Management Fund (SMF) grants. Government funding for waste initiatives has mainly come through the SMF and over 53 projects related to solid waste were supported between 1994-2003. An indication of the support for waste activities is clear when this is compared to the twenty-two projects funded by the SMF on wastewater and seven on air quality during the same period (Seadon and Stone 2003). Between 1994 and 2005, the Sustainable Management Fund provided $9.4 million in funding for waste minimization and waste management projects. Projects which focused on community education for wider sustainable development outcomes, but which also included waste minimization or materials recovery elements, also received an additional $5.4 million in funding. In total, funding for waste-related projects equated to around 36 per cent of the total fund allocation over that period. In essence there is a clear preference at the national level for voluntary agreements, educational programmes and partnerships rather than more interventionist strategies such as taxation.

In the past decade, there has been a move towards larger, privately owned landfills (especially in the North Island regions of Auckland and Waikato) or private/public partnerships, as in South Island communities of Canterbury and Southland. The costs of disposing to landfill have increased as standards of landfill management have improved and as facilities move towards full cost accounting for landfill disposal. However, despite pressure from private sector interests and environmental groups, the option of developing a national landfill levy, to internalize the unpriced costs of landfilling and provide an incentive to support waste recovery, has not progressed. The argument from central government is that improvements are being made by businesses and communities without having to resort to more national regulation (MfE 2006b). There is less resistance to interventionist strategies being developed at local levels, however, with the MfE supporting the introduction of local waste levies and pay-by-use regimes to cover the full cost of landfill disposal. In contrast to the view that national levies are unnecessary the MfE identifies as a highlight amongst significant achievements the amendments to the Local Government Act that allow territorial authorities to enact by-laws to set levies to cover any costs incurred in the administration of these functions and to fund waste minimization initiatives (MfE 2005).

The overall picture of waste management in New Zealand is then presented positively by the MfE with the conclusion that

> achieving excellence in waste management requires robust planning, community participation and innovation. As a nation, we can be proud of our achievements in waste management to date, and use them to inspire us to make further gains (MfE 2005, 2).

Challenges

While national government reports positively on the impact of legislative developments there are other commentators who are less impressed with activities in the waste field. Most outspoken in this regard was a review of plan making for sustainability, which concluded that 'New Zealand's brave new world under the RMA has not eventuated' (Ericksen et al. 2004, 283). The aspirations of the reforms, the authors claim, have been constrained by a combination of factors including a failure of governance at each level in the intergovernmental partnership fuelled by a lack of resources to implement new mandates at regional and local levels. As a result there has been controversy over the plans produced by sub-national governments, public discontent about administration and business complaints about compliance costs, while environmentalists express concerns about the focus on effects rather than creative pro-active planning. At centre stage of these criticisms is the view that national government has not provided sufficient direction or capacity at regional and local levels leaving authorities with dispersed populations and a low ratings base hard pressed to provide sufficient resources for waste management planning. While the intention of reforms was to give local authorities the space to innovate and find locally appropriate solutions the approach has tended to work best where local authorities already had good access to resources and operated under conditions of positive intergovernmental relationships. In its defence the MfE has itself experienced resource cuts following the enactment of the RMA leaving little internal capacity to support co-operative partnerships between regional and local councils, to generate inclusive public participation or to ensure quality through monitoring. In recent years a restructuring of the Ministry of Environment and the dissolution of the waste working group has also reduced the visibility of actions on waste management and except for the recent review of the Packaging Accord in 2004 there has been little movement on the key priorities identified in the Waste Strategy.

The devolution of waste activities both to local levels and to a wide variety of state and non-state actors has meant that information about waste flows in New Zealand has been fragmented and dispersed. In particular the high level of involvement by the private sector has led to an inaccessibility of waste statistics on the grounds of commercial sensitivity. There has been some progression since 1995 in this area with the National Waste Data Report, published in 1997, being the first aggregation of waste data at a national level and a Solid Waste Analysis Protocol was established in 2002. However no further national waste data reports have been published and solid waste performance indicators, while developed, have not been adopted nationwide (MfE 2005, 10).

Accurate information on volumes of waste is not the only problematic area for waste governance. There is also the complexity of waste flows where, for example, recyclables collected by a community group under contract to a council may pass through several companies, and across administrative boundaries, before final export or processing. Such flows of waste across space are also on the increase as landfills become consolidated into large scale regional facilities. As will be illustrated in the next section this geographical flow of waste can add to conflicts over the siting and management of landfills and other waste facilities. As a result of informational constraints and the fluidity of waste within a highly privatized system local authorities can exercise only limited direct control over waste streams.

The limited usefulness of using the RMA as a tool for up-stream mechanisms for dealing with waste is also problematic. It is not clear, for example, how a focus on effects can accommodate concerns with waste minimization. There is increasing pressure from some quarters to reframe waste management issues in order to focus on resource stewardship (Stone 2003; ZWNZ 2003) and the Māori concept of kaitiakitanga (stewardship). While the language of resource stewardship is present in the New Zealand Waste Strategy (MfE 2002) and other initiatives (WasteMinz 2001) there is concern that the rhetoric has not led to desired outcomes at the national level. Central government in New Zealand has been reticent in adopting legal and economic instruments to encourage waste minimization related activities and while the RMA provides overarching legislation for resource stewardship it does not have the specificities within its architecture to stimulate waste minimization (Stone 2003). Under the RMA wastes are only considered when they are have a negative impact on the environment. Their potential as resources to be managed is not addressed. As Stone suggests 'it is effects, not wastes that are required to be avoided, and there is still a choice of remedying and mitigating the effects they cause' (2003c, 14).

In terms of economic instruments there is no national landfill levy or plastic bag levy within New Zealand. Although several local authorities (including Christchurch and the North Shore are of Auckland) developed by-laws to collect waste taxes, on landfilled waste and to permit additional payments for usage of household collection services for non-recyclable materials, private sector companies challenged the validity of these moves and in 2006 their challenges were successfully upheld in the High Court.

The spatial signature of waste management in New Zealand is then constructed predominantly in relation to local conditions and politics and attention to developments at this scale is useful in illustrating the interactions between different waste actors. The final section of this chapter examines one particular development, a proposal to build a new landfill at Hampton Downs, within the Waikato Region of the North Island. It is a case that illustrates how the various actors come together through waste planning and RMA processes to influence waste governance outcomes.

Hampton Downs Landfill – the RMA in Action

During the 1990s it was becoming clear that many old landfills throughout the North and South Islands were either creating environmental problems or reaching capacity, or indeed both. With the shift in legislative frameworks through the RMA impacts of existing and proposed landfills were increasingly to the fore of discussions. A proposal was made to construct a landfill with a 30 million tonnes[3] capacity that would accommodate 24 million tonnes of waste providing a minimum life span of 35 years for Auckland and Waikato Regions. A site was selected at Hampton Downs in Meremere, a rural site located between Hamilton and Auckland and within the Waikato Region of the North Island. At the time it was the largest landfill, either established or proposed, in Australasia. The proponents already owned and leased the land as farmland and few people lived on the site and surrounding areas. The developers argued that the site, through its proximity to the State Highway, geology and groundwater containment, was suitable for landfill activities and proposed a restoration package that would increase the productivity of the land. However there was significant opposition to the proposal from local residents and environmental groups.

In March 1999 an application was made by Envirowaste Services Ltd and Northern Disposal Systems for twelve consent applications to construct the landfill. A joint hearings committee was set up between the regional and territorial local authorities Environment Waikato and Waikato District Council and consultants reviewed the application examining the technical and health assessments provided. During the hearing evidence was provided by 58 witnesses who were examined over a period of 79 days.

Ten consents relating to discharges to land, air and water, diversion of streams and water extraction, fell under the remit of Environment Waikato, the Regional Authority for the area. Two further consents concerning the transition of rural land to a landfill and in relation to upgrading the Hampton Downs Road were allocated to Waikato District Council. A joint hearings committee was set up in order to allow integrated consideration of the twelve consent applications. Initially both local authorities requested further information on technical issues and then a notification of the application was made in three local and regional papers as well as the New Zealand Herald. Due to the size of the proposal an extended statutory submission period was established.

Environment Waikato received 299 valid submissions, four supporting the proposal, six neutral, 14 of no opinion and 271 against, while Waikato District

Council received 321 valid submissions, seven of which were neutral and four supportive. Included in those supporting the proposal was neighbouring Hawaki District Council, Meremere Dragway Inc. (a waste collector) and Waste Management New Zealand (a major competitor of Envirowaste).[3] Neutral submissions sought further information and clarification regarding the type of waste being landfilled and on the processes to be implemented to ensure that nearby properties would not be affected by the development. Of the submissions opposing the application 121 were made on standard one page submission forms from people or organizations living or operating in nearby areas who were worried about land values, pollution, urban to rural waste transferral, transport impacts and the need for regular, independent monitoring if consents were issued. In particular there were concerns voiced about the lack of consideration of alternative locations or waste management methods. Ninety three more lengthy submissions, mostly from people or organizations from outside the immediate area, were also lodged. The concerns expressed by these participants focused on broader issues such as landfill not supporting a sustainable waste strategy therefore contravening local waste plans. Another comment made by external participants was a call for a reconsideration of an earlier failed bid to build an incinerator on the nearby Meremere site. Requests were made by these objectors for a full greenhouse gas audit, a full health risk assessment, contingency planning for fire and earthquakes and consideration of the impacts the landfill could have on waste reduction and recycling initiatives.

Overall opponents lodged over thirteen different objections that ranged from problems with landfill gases, odour, dust, pests and vermin through inadequate consultation, anti-zero waste implications of the landfill to the risk that consent for the application might lead to further undesirable applications in a rural location. Organizations as diverse as Airways Corporation New Zealand, the Fish and Game Council, Land Air Water Association, and Olivine New Zealand (an incinerator company) called for a full independent peer review of the application. There were also objections made from the local Māori communities. The Ngati Nako Hapuu (a co-operative society) claimed the area had a traditional significance and that importation of rubbish would lead to a denigration of the land and the community.[4] They reserved the right to kaitiakitanga (stewardship) and felt that there had been insufficient attention paid to the impact of the development on local kai (food)

---

3   It may seem surprising that a major competitor of Envirowaste would support the proposal but at the time a legal case was pending relating to a possible case of monopoly with respect to Waste Management New Zealand and other actors involved in the case felt that it was in the company's interest to get Hampton Downs approved.

4   Māori are specifically identified as being important actors in calling for improved environmental standards including waste based on their customary practice and strong sense of duty to protect the environment as Kaitiaki (stewards). The relationship of Māori to the environment generally is complex and has been discussed elsewhere (Awatere 2003; Patterson 1992; Pawson and Brooking 2002), but fundamentally the issue with waste is to ensure that it is being disposed of appropriately, without damaging the environment that sustains the tangata whenua (people of the land). Inappropriate dumping of waste into mahinga kai (food-gathering areas) can degrade the mauri (the life essence of any living creature or thing) and the food producing capability of the land.

resources and inadequate consultation with the local hapuu (sub-tribe, extended family). In particular they opposed the discharge of pollutants and the taking of water from streams and tributaries. Waha Pu O Waikato, representing a number of Waikato hapuu, claimed the landfill was planned on confiscated lands and identified a strong tribal interest in the area.

In response to these objections Envirowaste asserted its position as a waste management rather than a landfill company and suggested alternative sites had been considered through a selection process that occurred over a two year period. They also considered that the greenhouse gas emissions produced from the site were not at nationally significant levels and that their facility was actually a storage site for potential emissions. Envirowaste had established a peer review panel, at its own cost, to evaluate the proposal and had provided assurances that an aftercare plan would be established before the landfill was closed. Both Waikato District Council and Environment Waikato had required a bond to cover planting and landscaping provisions. In addition the applicant dismissed objectors concerns regarding the developments impact on the landscape on the grounds that the area was not identified as outstanding or significant in either District or Regional plans. With respect to the issue of the landfill development contravening local plans Envirowaste pointed out that the Waikato District Waste Management Plan (1999) did not rule out landfill development, only landfill development constructed and run by the council.

In contrast to the claims made about inadequate consultation Envirowaste listed a whole set of meetings with immediate neighbours and landowners, site visits for interested parties to other facilities owned by the company and information provision. Envirowaste said it had produced regular community newsletters and project discussion documents as well as holding open days and personal consultation with Tangata Whenua, environmental groups and regulatory authorities. These processes had occurred at the project announcement, during the release of the concept plan and finally with the production of the landfill plans. In response to these procedures Envirowaste had purchased land from three nearby property owners where traffic effects were potentially detrimental, proposed a new access road and acoustic bunding to reduce noise impact and had developed a proposal to transport leachate to Mangere Treatment Plant in Auckland. They did acknowledge however that despite these changes some legitimate concerns remained.

At the joint hearing Waikato District Council, while raising some concerns about locating a landfill on an area with high quality soils, concluded that the adverse effects on the environment from this facility would be minor and that it was consistent with the RMA. They recommended consent be given subject to seven conditions and a timescale for the development of seven years. In addition to general conditions relating to compliance with plans the consent was given on the proviso that no hazardous, special medical or radioactive waste would be dumped at the landfill and that random checks for non-permitted waste would be conducted. It was also made clear that all costs for review following changes to standards in relation to landfills were to be borne by the holder of the consents. The consents would be reviewed after one year and then every two years subsequently, with the consent holder able to apply at any time to change or cancel conditions. In addition Envirowaste was to offer local residents an opportunity to establish a liaison group that would involve a two-monthly site inspection, provisions for reasonable information and 24 hour expertise to deal with public complaints.

The conclusion of the Officers Decision Report noted that that the hearing was not the appropriate forum to promote alternative systems of dealing with waste and that there was a clear misunderstanding by some objectors of the role of Environment Waikato in the development. The report reiterated that Environment Waikato was not involved in the management or ownership of landfills rather it was responsible for controlling and regulating environmental effects of waste management practices. Despite this the decision letter did support concerns about the amount of recyclable or re-useable material that ended up in landfills, but suggested that while waste was still being produced it was necessary to dispose of it.

Objectors to the development maintained their opposition to the facility and an appeal was made to the Environment Court. The Waikato Times reported that in July 2000 protestors from Meremere organized a 34 km hikoi (march) from Te Kauwata to Ngaruwahia to present their grievances to Waikato District Council. In 2001 the New Zealand Green Party also entered the debate with a statement from local Green MP and spokesperson for the Waikato Greens, Nandor Tanczos, rejecting the proposal for a landfill. While recognising the need for landfill capacity, Nandor resolved that the party would need to be satisfied that any proposed waste disposal projects are environmentally safe, culturally acceptable, and part of a longer-term strategy of waste minimization. He particularly made reference to the cultural and historical significance of the site, noting that eight iwi and hapuu groups had joined the appeal against the facility. He agreed that the Waikato River was of spiritual importance to the hapuu of Tainui playing a key role in Māori history and everyday life.

The Environment Court reconsidered the proposal for the landfill over a ten month period which was reconfigured as a case between Waikato Regional and District Council and the appellants. The appellants included two local residents and two local environmental associations. In contrast to the initial joint hearing held by the local authorities the Environment Court operates under legal conditions with a formal swearing in of participants. The hearing followed a series of judicial conferences when the Court issued orders, directions and minutes in an endeavour to narrow the focus of the parties and Court's attention on the contestable issues. However the appeal was fought over a broad front with appellants electing to challenge the evidence of almost every witness, much to the annoyance of the Judge. The main areas of contention were: whether the proposal was contrary to relevant statutory instruments, whether there were adverse potential effects on the environment and whether Māori cultural, spiritual and consultation issues had been dealt with adequately.

In relation to procedural issues the appellants claimed that the location of the landfill close to the Waikato River was contrary to policies in the District Plan to maintain the amenity of the river and protect high quality soils, rural visual character and amenity values. However the wording of the plan had been changed to allow discretionary use of rural land for solid waste management activities. As a result the Judge concluded that if the landfill was constructed and operated according to the conditions established by the local authorities it would then be in accordance with the plan (Environmental Court 2001). The appellants then raised Regional Policy

Statement Sec 3.9 of the Regional Council Plan that seeks good management of waste according to the waste hierarchy through

> The efficient use of resources, reduction in the quantities of wastes requiring disposal in the Waikato Region, and the adverse effects associated with their generation and disposal.

In this case the judge rejected the submission out of hand as irrelevant to the impact of the facility under consideration, stating that

> The extent to which a local authority should become involved in landfills is a matter outside the ambit of these proceedings. Envirowaste having made an application for resource consents is entitled to have its application heard on its merits (Environment Court 2001, 22).

As well as bringing inappropriate challenges the Judge also concluded that the quality of the appellant's evidence was not always up to standard. In one situation where issues relating to potential risk and health concerns the Judge concluded that

> overall we prefer the evidence of Mr Kennedy (for the developers). Not only is he more experienced than Ms Bell (for the appellants), but we consider Ms Bell's evidence to lack the objectivity we would expect from an expert witness (Environment Court 2001, 354).

While the Judge acknowledged that there were gaps in understanding of the different ways that chemicals can affect health he supported the standard risk assessment protocol used by the developers in the case. He noted that

> it is not possible, nor is it feasible, to estimate the potential risk associated with all chemicals and mixtures that might be disposed of in a landfill. Using a well-accepted generic list of volatile chemicals for the most significant exposure pathway is a reasonable and widely accepted approach, though not one approved by [the appellant] (Environment Court 2001, 59).

The issue of judging between the findings of competing expert witnesses was also raised in relation to the treatment of Māori cultural issues that the proposal impacted upon, such as the discharge of effluent to the river and the extraction of water to manage the landfill site. On the one hand Dr. Michael King who wrote a biography for the region *Te Puea* (1977) made the case that

> more than any others in New Zealand, the tribes of the Waikato Valley are a river people. Five centuries of continuous occupation of its banks have embedded the river deep into the group and individual consciousness. More than any other gesture, living alongside the river was an affirmation for Waikato people of who they were (Environment Court, 2001:101).

Similarly other Māori support for the protection of the river's integrity stated that the river is regarded as out ancestor, whereas modern developments, hydropower and effluent disposal, are regarded as an affront and deliberate desecration of our tribal ancestor (taonga) (Mr Falwasser, recorded in Environment Court 200, 101).

However there was also a Māori, Mr Mikaere, speaking for the defendants who challenged the sacredness of the site that had been justified with unnamed individuals and on the basis of anecdotes rather than recognized sources.

A major issue for debate, and one that has been present since the Treaty of Waitangi was signed, was the translation of Māori terms and definitions of their meaning into the English language. It was established that it is one thing for a Māori to give evidence in terms of their customs and quite another thing again to give evidence that explains them. Appellants felt that the Court lacked expertise in tikanga Māori (things Māori) and therefore was not able to appreciate the nuances of the language used. This was rejected by the Judge who stated that on the contrary, despite not having such expertise the Court was able to make a determination on the evidence, just as it has to make determinations on many matters which are outside the professional expertise of its members. In relation to making that determination it was acknowledged that the construction of the landfill, by its very presence diminishes the mauri (that power which permits these living things to exist within their own realm and sphere), but that a decision had to be made in the context of the effects of previous pastoral development and the artificial channelling of streams for agricultural purposes and the unlikely contamination through discharge of the water bodies. In summing up the Judge stated that while

> it had been understood well enough that cases involving Māori values require individual consideration and assessment, without there being any overriding presumption that tangata whenua (people of the land) may effectively veto a proposal. Issues of waahi tapu (sacred site) and the like require to be weighed and determined objectively in the circumstances of the particular case, without allowing the pressure of concerted and sustained opposition to achieve a predominant influence and deter an appropriate outcome consistent with the Act's overall purpose (Environment Court 2001, 107).

When comparing the cases made by the witnesses for appellant and defendant the Judge erred on the side of the defendants for, while recognising that the landfill site occupies ancestral lands of importance to certain hapuu, he claimed that

> the evidence put forward in support of the assertion [that the land is a sacred site] was either hearsay, or general in nature and lacked any specificity by way of oral tradition or historic foundation. This is in contrast to the carefully researcher and well-reasoned evidence of Mr Mikaere supported by Dr Clough. We prefer and accept their evidence. The site is not of any particular cultural significance (Environment Court 2001, 119).

The judge also expressed his concerns that the tangata whenua representation was only brought up at appeal and not two years previously at the Council hearing. In his summing up the Judge was damming in his synopsis of the appellants case stating that while we have no doubt that those opposed to the proposal are sincere. However, it is unfortunate that through a lack of focus the combined opposition of the parties and the presentation of their cases, was disjointed and inchoate (Environment Court 2001, 26).

It was the Court's decision that

the landfill as regulated by the conditions of consent, will conform with the relevant objectives, policies and rules of the relevant statutory instruments ... we conclude that granting consent subject to the conditions attached will promote the sustainable management of nature and physical resources as defined in the Act. Consent should be granted accordingly (Environment Court 2001, 145).

Not only did the Court's decision go against the appellants, but unusually the Judge awarded the appellants costs, which was seen as a punishment for prolonging the process.

Following the decision there was much consternation, unsurprisingly from the appellants but also from the defendants. Interviews where held with public, private and civil society actors involved in this case in order to examine the reasons behind the dissatisfaction and illuminate the practices of governance that lie behind the RMA process in relation to the governing of waste. A major grievance of the appellants was that the proposal, due to its scale, was of national significance and a national hearing should have been held rather than the joint hearing between Environment Waikato and Waikato District Council. Allied to this were concerns about the lack of investigation of other options for waste disposal and the lack of detail and definition in local waste management plans. As one appellant stated

the Waikato District Waste Plan says they are committed to reducing the amount of waste to landfill. How is building the largest dump and importing waste into the area reducing waste to landfill? (Civil society 3).

The fact that the RMA does not permit such dimensions to enter debates about particular developments was seen as a fracture between policy structures, between the waste plans and the RMA process. Citing the subsequent approval given to a correction centre and a motorsports facility in the Hampton Downs area concerns were raised that attention to impacts of discrete developments allowed the gradual destruction of environmental quality, or as one appellant put it, leading to 'a case of death by a million cuts'. Rather than working in concert towards a common goal of sustainable development the waste planning process and the RMA consenting process were seen to be at odds with each other.

A second area of concern raised by appellants related to perceived imbalances of power between participants that were exacerbated by the structure of RMA procedures, particularly once proceedings had moved to the Environment Court. In the words of one appellant

the RMA is a well written document, but it can be manipulated by lawyers and it can't be used correctly if unequal funds exist between developer and appellant. It's not an even playing field and it's not working (Civil society 4).

Another appellant found themselves without legal representation just before the case was heard in the Environment Court and felt that

> I should have had legal aid. The hearing should have been adjourned ... it was up to the judge to do this, a lay person couldn't know ... I was in a weak position and as a lay person I used the wrong terminology (Civil society 2).

The developer in the case, while happy with the overall outcome, also expressed concerns about the costs of the RMA process and the conditions attached to the consents approved. Even before Hampton Downs was opened over $21 million NZ had been spent by the developer and in part it was felt that these costs were the result of the devolution of responsibility to sub-national scales of government. Such devolution was seen as problematic given the differential levels of expertise and resources within local authorities. Although for different reasons the developers, as with the appellants, felt that there was not a level playing field in RMA based decisions as a result. In the Hampton Downs case the developers felt that the local authorities were being over cautious because of their responsibility for supporting the proposal.

> It's related to Environment Waikato's interpretation of the RMA. It's belt and braces. It's way overboard, way too engineered because of local opposition ... if anything ever goes wrong the local people would just kill Environment Waikato so they dumped all of that responsibility onto us and trebled it ... not only are the consents different in every region, but the interpretation and enforcement is different too (Private sector 3).

However while there was some support for stronger consistency nationally in terms of interpretation and enforcement of the RMA the developer did not support the views of the appellants that the Hampton Downs case should have been referred to the Ministry for the Environment. Despite the numerous individuals and Members of Parliament writing to the Minister for the application to be called in the developer asserted that 'it's got nothing to do with the Ministry, it's the regional council's responsibility ... there was no way it was a nationally significant issue'. There was then a mismatch between the appellants who felt that the scale of the landfill and the seriousness of waste as an issue meant that it was nationally significant and the developers, who were focused on the requirements of the RMA, only saw the facilities impact on waterways as relevant to the debates. As one interviewee stated when asked what categorized an issue as being of national importance

> effectively [to be nationally significant] it has to have something like climate change involved and say if someone is proposing to build a coal fired power station. Definitely not in this case which was a local waterway issue (Private sector 4).

At the heart of the disagreement over the significance of the case is the fact that waste per se is not managed as an issue through the RMA. RMA consents are related to emissions and discharges. Yet, as one local authority officer suggested, problems arise because waste management plans are often broad in focus and there is no forum to discuss wider waste issues except at the resource consent stage. He commented that

Stronger plans would have the ability to be more controlling of disposal, but it is not going that way. The MfE is not becoming more active or controlling so we have the problem that the resource consent stage is where we have the debate (Local public sector 5).

Despite recognising this tension local authority officers did feel that the Hampton Downs case had been unnecessarily prolonged by appellants, costing Environment Waikato $250,000 NZ. While acknowledging that there was variation in application of RMA consents at a national level the local authority officers did not feel that their judgements had been overly constraining and that their in-house skills allowed them to provide a solid basis for decisions and enforcement.

The Hampton Downs case was a particularly high profile case of the difficulties of governing waste through the RMA. It highlights the different perspectives of public, private and civil society actors operating through a process that is deliberately scaled to focus decisions on to specific developments occurring in particular localities. The comments of interviewees involved in the case reiterate general perspectives that, despite the emergence of a national strategy, waste as an issue is not strongly regulated at a national level and not considered holistically through the current structures of the RMA. However, as mentioned in previous sections, the RMA is not the only mechanism that can be used to govern waste and the final section of this chapter considers the outcomes of overall waste management system in New Zealand.

Conclusion

Since the early 1990s New Zealand has seen the emergence of a more sophisticated waste management system driven by political, environmental, economic and social factors. Initially focused on improving established mechanisms for disposing of waste attention has also turned to issues of waste minimization stimulated particularly by increased awareness of innovative tools to reduce waste generated and by the increasingly visible activities of the non-governmental sector in New Zealand. While it seems that the dominance of landfill is likely to continue, with permission being granted for large regional landfills in both North and South Islands, the lobby for resource stewardship is still active and Green Party MP, Nandor Tanczos, has developed a Waste Minimization Solids Bill that is being considered by a government select committee.

Overall critics remain concerned about a lack of co-ordination in waste prevention and minimization, uneven attention to waste reduction and recycling amongst local councils and variable levels of information, knowledge and expertise across the country. They also call for greater leadership by national government through a central agency responsible for co-ordinating waste prevention and minimization initiatives through mandatory legislation. The private sector, as the dominant waste service provider, feels overburdened with demonstrating environmental effects of its activities whereas civil society actors call for more assistance to participate equally within the RMA process. In particular commentators have expressed concern about the movement of waste management issues out of the public sphere to unaccountable corporations, subject to the vagaries of market forces and unelected judges or scientists.

PART 3
Comparisons and Conclusions

Chapter 7

Comparing Garbage Governance: Shades of Green Governance

Comparing Country Overviews

To date waste governance analyses have tended to take one of two approaches. The first is to examine in detail the processes of governance in one country (e.g. Bulkeley et al. 2007), the second compares certain aspects of waste governance such as waste production, policy instruments or institutional structures in widely differing contexts (such as Hill et al. 2002 and Parto 2005). The aim here is to combine the benefits of both approaches and undertake a detailed examination of waste governance – from policy structures, through interactions between spheres and tiers of governance, to policy outcomes – in two countries that exhibit broadly similar characteristics in terms of population size, history and development. More specifically both countries have experienced a growth in waste production in recent years and have adopted the waste hierarchy as the overarching framework for sustainable waste management.

The focus on similarities between the two countries should not be overemphasized however as it will be seen that commonalities co-exist with significant differences and even apparent similarities can conceal diverse experiences. The first section of this chapter compares the histories of the two countries in terms of their political, economic and cultural evolution before looking in detail at the governing systems that are employed, the interactions between governing actors and agencies and finally the outcomes of those governing experiences.

Political and Economic Development

Historically a unifying feature of Ireland and New Zealand's political and economic development is their common colonization by the United Kingdom. Although both countries are now independent New Zealand remains part of the British Commonwealth and retains Queen Elizabeth as its head of state. Each country has developed a parliamentary democracy, but beyond this the nature of inter-governmental structures diverge. There are two elements of this divergence that merit particular attention. The first major difference is that Ireland operates within the wider supra-national governing structure of the EU, a level of government that is clearly absent in New Zealand, and the second is the allocation of responsibilities to sub-national governments. New Zealand operates a devolved system of governance that gives regional and sub-regional (territorial) government significant powers while in Ireland local government remains relatively weak and regional structures virtually absent from daily governing activities.

Both countries have a population of just over four million, but the larger landmasses that make up New Zealand mean that it has a relatively lower population density and, following the recent economic boom, Ireland exhibits a slightly higher rate of population increase. It is the nature and extent of Ireland's economic growth that provides a key difference between the two nations. For while historically both countries have relied heavily on agriculture as a node of economic development, and both experienced periods of economic recession during the 1980s, Ireland's emergence as a centre for the hi-tech sector in the 1990s led commentators to consider whether New Zealand should look to become the Ireland of the South Pacific (Bollard and Box 1999; Box 1998). Analysts pointed to the investment incentives, wage agreements between unions, employers and government, special low tax rates and a highly skilled (and growing) workforce that had attracted inward foreign investment to Ireland. Although these internal strategies undoubtedly assisted in economic development external factors were also significant. Ireland benefited from economic stability through the European Exchange Rate Mechanism (ERM) as well as from EU structural funds and agricultural support. Ireland also had the geographical advantage of being an English-speaking nation on the edge of Europe whilst also having historical trading links with the USA. Nevertheless despite these divergent economic pathways both countries still rely heavily on the intersection of nature and society for tourist activities.

Nature, Culture and Society

While financially beneficial the recent economic growth experienced in Ireland also created significant pressures on the environment that brought to the fore discussions about relationships between nature, culture and society. In contrast such discussions had long been a feature in New Zealand following on from contestations over stewardship of the environment raised in the Treaty of Waitangi and through subsequent resource exploitation by settlers. Both countries had, however, experienced past periods of humanly induced environmental change. In Ireland changes had occurred since the Neolithic period with a peak in impact following population expansion in the late 18th century. While the settlement of New Zealand was a more recent affair, stretching back only around 800 years, it was again during the 18th century that major landscape changes occurred with mass forest and scrub clearance.

The periods of environmental degradation experienced in both countries during the 18th century were clearly influenced by colonial factors and it has been suggested that these experiences also played a role in shaping societal attitudes towards nature and the environment. In New Zealand the relatively rapid development of natural resources led to strong reactions from both Māori populations and sections of Pākehā society concerned about overexploitation that eventually led to the development of a strong environmentalist movement and the first green party. It also contributed towards development of an environmental protection regime during the early 1990s that gained global acclaim for its foresight. Although the impact of this legislation has been contested the idea of protecting New Zealand's natural environment and the nation's 'clean, green' image, retains a revered place in the national psyche.

In contrast the devastating impacts of crop failures in Ireland during the mid 1800s have been linked to a conception of environmental management as a concern for elites. Land is highly valued across the Irish population, particularly in rural locations, however it has tended to be defined more in terms of utilitarian value and private property rights than in terms of nature conservation or environmental protection. While recent surveys of public attitudes towards the environment in Ireland have indicated a convergence of concern for the environment with European counterparts, environmental practices and activism have yet to become mainstreamed. Nonetheless despite the absence of a strong national environmental movement in Ireland a structured system of environmental regulation driven by external pressure from the EU has developed.

Environmental Policy

As with the comparisons of political, economic, social and cultural contexts environmental policy evolution in Ireland and New Zealand demonstrates a mix of commonality and differentiation. Most significantly while both countries have enhanced the sophistication of regulatory frameworks and have adopted the common goal of sustainable development as a guiding rhetorical framework the regulatory mechanisms adopted suggest different policy trajectories. The key here is that the environmental policy regimes in both countries were not constructed in isolation from either the social and political changes or cultural and economic contexts detailed above.

The major feature of New Zealand's environmental policy landscape has been the formulation of an effects-based strategy, the 1991 RMA, within the context of broader neo-liberal restructuring of governing systems. Although the Ministry for the Environment provides broad guidelines for environmental policy, and the Parliamentary Commissioner for the Environment exists to monitor the quality of environmental decision making, day-to-day resource management decisions were devolved to sub-national tiers of governments through a system of plans and policy statements in order to facilitate a liberal regime for developers. Wider participation in decision making was facilitated through opportunities for submissions on developments from any interested party whether they were supportive, opposed or neutral in their assessment of the proposal. The RMA also included channels for appeal against the decisions made by sub-national governments although this process was tightly formalized with strict timelines and legalistic protocol through the Environment Court. The aim of the RMA was to restrict evaluations of environmental effects to legal and scientific frameworks rather than broader environmental, ethical, social or political perspectives. However such an aim assumes the existence of universal environmental standards and precise understanding of biophysical processes. The success of the system then depends on whether sub-national governments have the capacity to determine, consistently and accurately, the effects of a proposed development on the environment. In essence what emerged in New Zealand was an environmental regulatory regime that sought to marry reform of government based on neo-liberal principles with a resource management process that would satisfy a concerned population. Powers were devolved to sub-national government and the courts while decisions about environmental impacts were reduced to scientific judgements about effects of developments.

Although the RMA was heralded as a progressive environmental strategy at the time of its publication subsequent analysis of New Zealand's movement towards sustainable development, and the RMA's contribution to that goal, have not been so positive. In 2002 the Parliamentary Commissioner for the Environment concluded that 'New Zealand could have been a leading light on sustainable development by now, but we are not' (PCE 2002, 1). In particular concerns were raised about a lack of national leadership and co-ordination in terms of forward planning for sustainable development. In response a programme for action on sustainable development was published in 2003 to supplement the 1995 sustainable development strategy. A similar pattern can be identified in Ireland where an initial strategy was 1997 and a review of progress conducted for the Johannesburg Earth Summit in 2002. In both cases the sustainable development strategies make general statements about policy developments and outcomes and both articulate a desire for a 'win-win' situation in terms of economic growth and environmental protection. Essentially both exhibit the rhetoric associated with ecological modernization.

In contrast to New Zealand, Ireland's environmental policy has been defined by the formation of a single, dedicated environmental protection agency following the 1992 Environmental Protection Agency Act. As with the formation of the RMA in New Zealand, the primary motivation behind regulatory reform was to consolidate environmental decision making and elevate scientific assessments of environmental quality. However, rather than create an overarching act to control resource management, it was decided that a new centralized agency would more effectively overcome the limitations of the current regimen criticized for being ad hoc, reactive and clientilist. A major consideration for Ireland was to develop a system that could accommodate the increasing number and demands of EU Directives while facilitating economic development. The formation of an environmental agency created a clear competent authority, for EU purposes, responsible for issuing integrated pollution control licenses, providing support for local authorities through environmental reporting and enforcing environmental legislation. However the EPA did not take over all roles associated with environmental planning and local authorities remain responsible for issues such as traffic, landscape and visual impacts through the land use planning system. For example, if a landfill is proposed the developer will need to gain a waste license from the EPA and planning permission from the relevant planning authority. One issue with having a parallel process is that the delineation of responsibilities between the agencies may not always clear cut and each system has its own processes of evaluation and consultation. In relation to public participation, for example, in decisions about developments there are statutory periods of consultation in the land use planning system, but the EPA decides whether or not to hold an oral hearing based on a judgement about the existence of a scientifically valid objection.

In terms of environmental policy evolution then the key difference between the two countries is the location of decision making and enforcement powers. In New Zealand central agencies have taken a back seat while sub-national local authorities make decisions (and set conditions) on resource consent matters using a legal system to deal with any cases of appeal. In Ireland a national agency, the EPA, takes centre stage in relation to pollution control and enforcement. Recent changes to planning procedures have also served to centralize decision making over strategic infrastructural developments, which are now dealt with directly by the national planning appeals board leaving little room for local authority influence in decision making.

Waste Policy

As indicated above in relation to discussions of sustainable development strategies environmental decision making is not restricted to development control issues and there have been significant changes in both countries in relation to forward planning within specific environmental sectors. In relation to waste the commonality of experiences between Ireland and New Zealand up until the 1990s are quite marked. Both countries were reliant on a system of local landfills, often without environmental controls, for waste disposal, levels of recycling were low and no municipal solid waste incinerators were in existence. In both countries the ad hoc nature of public health and local government acts used to manage increasing volumes of waste were increasingly recognized as deficient and demands for a more holistic approach to the treatment of waste matters were developing.

New Zealand began by setting explicit targets for recycling in 1990 although these were subsequently dropped when progress towards attainment was slow and the Waste Management Policy published in 1992 instead emphasized the importance of attention to the waste management hierarchy in waste programmes. The waste management hierarchy also came to the fore in Ireland through developments within the EU and by 1996 both countries had begun to institutionalize more sophisticated waste management planning regimes. In New Zealand this was achieved through an amendment to the Local Government Act while a specific Waste Management Act was developed in Ireland. Both systems called for the production of sub-national waste management plans for municipal solid waste. There was however more prescription regarding the waste planning process in Ireland and the legislation provided for more central control over the nature and form of plans. Regionalization was supported with the Minister for the Environment able to require local authorities to co-operate through regional plans enshrined within the Act. Enforcement of this capability was not necessary in practice however as the majority of local authorities chose to regionalize voluntarily on the grounds of efficiency and economies of scale. Similar calls for regionalization of waste plans are emerging in New Zealand and a few such plans have recently been developed, but overall regional or local waste plans lack the depth and detail of their Irish counterparts.

The dictates of Europe are crucial as the detail in Irish waste plans stems primarily from the need to address the demands of the 1999 EU Directive to divert waste from landfill within prescribed timescales. This influence can also be seen in relation to the collection of waste statistics where the onus is on Ireland to demonstrate compliance with the diversion targets set within various waste directives. While both countries recognized the need for regular, accurate information on waste, only Ireland has institutionalized practices to achieve this, collating six reports on waste statistics since 1998 compared to the single national waste database survey conducted in New Zealand in 1997.

Despite the absence of a supra-national government directing attention in the waste arena there was considerable internal pressure, particularly from environmental and community groups, to address the growing waste problems in New Zealand that culminated in the 2002 National Waste Strategy. Not only was this the first national strategy produced on waste but it also incorporated in its title a commitment to moving towards zero waste and sustainable development. Yet nowhere in the Strategy was there a clear outline of what waste zero waste might actually mean in practice nor was there a definition of zero waste provided in the glossary of the document. Far from the potentially radical interpretations provided by zero waste campaigners the Strategy notes that a

> zero waste and sustainable New Zealand requires new ways of thinking at every level of the community. It doesn't mean radical change – we don't have to avoid the products and services we normally use – but we do have to think smarter about the service we want from products and find better ways of getting it (MfE 2002,19).

Certainly the tendency for a strong lead in environmental matters from central government in Ireland stimulated by pressure from Europe is replicated in the waste field through the form and content of the Waste Management Act (and amendments). The Act incorporates a considerable degree of prescription both in terms of defining waste management planning for sub-national governments and in detailing specific targets for waste management practices. In New Zealand the production of a National Waste Strategy initially suggests a stronger lead from the centre in waste management matters than in other environmental sectors. However on deeper inspection the document does not propose radical amendments to existing Local Government Acts or the RMA in relation to waste management, local authorities are not provided with detailed frameworks for the production of plans and the paucity of accurate data on waste production and disposal prohibits the development of ambitious targets.

Interactions

In both countries the structuring forces of the policy interventions detailed above have emerged, and are constantly being reshaped, by interactions between different spheres and tiers of governance. Actors from private and civil society sectors, both indigenous and international, negotiate the detail of policy frameworks as well as implementation practices and enforcement regimes. The concern here is to examine the nature of relationships, or interactions, between the various spheres of governance in the two countries in order to establish areas of commonality and divergence.

In Ireland and New Zealand the private sector is a dominant actor in waste management service delivery. In Ireland only 40 per cent of municipal waste is publicly controlled with the remainder either sub-contracted to the private sector or a purely private affair. The figures are even more extreme in New Zealand with only 10 per cent of collections entirely owned and operated by the public sector, a clear corollary of neo-liberal reforms during the 1980s. Importantly however most of the private sector services in New Zealand are constrained by contracts with local authorities whereas only 10 per cent of private services in Ireland operate through sub-contracting. This means the local authorities in Ireland have less control over the ways in which waste is collected and disposed of. Yet these figures do not reveal the nature of private sector contributions. In both cases there has been a consolidation of the waste industry with the emergence of national, and even international, waste companies controlling large elements of the waste stream. In Ireland the private sector has also played a significant role in the formation of waste management plans. As detailed in chapter 5, following the 1996 Waste Management Act, local authorities were required to seek assistance from waste experts when drafting their plans. These waste experts were, without exception, drawn from the private sector. Equally environmental consultancies have been pivotal in developing and providing waste awareness initiatives both at a national scale (Race Against Waste) and sub-nationally (e.g. Dublin waste-to-energy project). Waste plans and waste awareness campaigns in New Zealand, in contrast, have been very much driven by local authorities with assistance from civil society groups.

The role of civil society in waste management issues is a clear area of divergence between the two countries. For while civil society activism is present both in Ireland and New Zealand, the level of activity and the extent of influence exerted by this sphere of governance in policy making circles differs widely. Ireland's waste related civil society organizations – be that community based recycling organizations, anti-incineration campaigns or organized protests surrounding waste charges – are few in number, weakly networked and generally marginalized from decision making circles. In contrast New Zealand has hundreds of community resource or recycling organizations that provide a whole range of services and contribute significantly to local waste management strategies in particular areas. These groups communicate and collaborate through a variety of networks, such as Zero Waste New Zealand and the Community Recycling Network, that provide a national voice for their concerns in policy debates. Key figures in these organizations were influential in ensuring that the concept of zero waste was included in the New Zealand waste strategy and continue to lobby governments to address waste reduction and minimization issues. What is interesting here is that despite the devolved system of government in New Zealand there exists a strong national base within the waste civil society sector. Yet in Ireland where the system remains more centralized there is an absence of such national organization. So why should this disparity exist? One reason for the divergence may be located in the histories of environmental activism within two different countries. It has been suggested that popular environmental protest in Ireland has been dominated by localized community responses to discrete environmental threats whereas there has been much more of a campaigning tradition focused on broad natural resource management issues in New Zealand. This attention to issues of resource stewardship may be linked to the influence of Māori culture (Kaitiatanga) as enshrined within policy making through the Treaty of Waitangi in combination with the particular nature-society relations that have emerged since the European settlement. Equally the lack of a strong nationwide environmental movement in Ireland may also be linked to particular conceptions of nature-society relations that have been coloured by past periods of colonialism.

Attention to the relative nature and form of private sector and civil society spheres of governance is important, but despite considerable neo-liberal reforms in New Zealand and trends towards privatization of the waste stream in Ireland it is still the case that the public sector retains the power of redirection in relation to policy formation, implementation and enforcement. How governments at different levels interact with the different spheres and tiers of wider governance remains pivotal in shaping governance outcomes. In the centralized Irish system the approach of national government is crucial here and it has generally been characterized as a developmental state (Boyle 2003). There is strong support for promoting technical solutions to environmental problems, including waste, and big business is seen as a trusted partner (albeit still in need of regulation) in service delivery. Environmental policy regimes are conceptualised as facilitators of development couched within rhetoric of ecological modernization. Evidence of similar rhetorical devices can be seen within environmental policy documents in New Zealand however the central state, particularly the Ministry for the Environment, is much less prescriptive than its Irish counterpart in terms of policy frameworks for waste management.

At the same time as being less prescriptive the Ministry for Environment in New Zealand is perceived as being more open to discussions with civil society organizations despite the partnership agreements that have characterized elements of policy making in Ireland in recent decades. Of course the openness of central government to civil society groups does not mean that they have a privileged position in either policy negotiation or implementation. Indeed it was a concern of civil society groups in New Zealand that they were seen as the same as the private sector when competing to attain waste management contracts with local authorities despite the added environmental and social benefits offered in their practices. In addition while the RMA incorporates opportunities for public participation it has been argued that the effects based focus and legalistic framework of the RMA actually serves to subtly oppress participation of those without institutional or corporate experience (Gunder and Mouat 2002) and such criticisms might equally be levelled at participatory mechanisms in an Irish context.

Waste governance landscapes in Ireland and New Zealand are then created by a complex of historical and cultural conditions that combine with contemporary interactions between actors and agencies from all spheres of governance at different scales. The form of these interactions is not set in stone, rather they are constantly being contested and the final section of this chapter considers the outcomes of these policy interventions and dynamic governance interactions.

Outcomes

Outcomes in the governance sense can be identified as the discourses or narratives that are developed to conceptualize waste, the policy mechanisms institutionalized to manage waste and more materially the impact that those discourses and mechanisms have on the nature and volumes of waste being produced.

In relation to conceptualizations of waste the dominant articulated narrative of governance actors in both Ireland and New Zealand is that of a waste management hierarchy. As previously detailed the hierarchy considers waste prevention and minimization as the ideal means through which to deal with the by-products of contemporary society, followed by reuse, recycling, waste-to-energy recovery and finally disposal. This common acceptance should not be read as discursive harmony when it comes to conceptualising waste management issues however. There are calls in New Zealand, for example, for the language of hierarchy to be replaced by a more integrated notion of resource stewardship (Stone 2003) and at the same time voices from within civil society are arguing for a more definitive move towards waste prevention and minimization through the language of zero waste (Connett and Sheehan 2001).

The influence of these zero waste activists can be clearly identified in the sub-clause of the title of the New Zealand Waste Strategy '*towards zero waste and sustainable development*', however what this concept might mean in practice is not addressed in the body of the document. In part this acknowledgement of the zero waste idea by central government must be linked to the achievements of Zero Waste New Zealand and funding provided by the Tindall Foundation. Together they provided funding and expertise to local authorities in support of zero waste. In 2001, before the Strategy was published, 40 per cent of local authorities in New Zealand had committed to achieve zero waste to landfill by 2015 or 2020 in their waste management plans. By 2006 this had increased to 70 per cent (51 out of 71 councils). To qualify as a zero waste council a minuted resolution at a full Council meeting has to be made confirming the council's commitment to a target of zero waste to landfill by 2015, with a review in 2010 (to allow Councils to re-evaluate the zero waste target in relation to its obligations under the Local Government Act, Amendment No.4). A commitment also has to be made to full and open community consultation and ownership of a zero waste strategy involving community, council and business sector partnerships.

Although a zero waste organization exists in Ireland (Zero Waste Alliance Ireland) it has yet to exert an influence over policy making, either at the national or local level, that compares to the experience of New Zealand. Indeed a past Minister for the Environment is recorded as saying

> those advocating a zero waste policy have zero credibility ... Had the Government adopted such an approach in our regional waste management plans, Ireland's waste management capability would be in a sorry state today, people would be paying more in taxes and foreign investors would not come to Ireland (DoEHLG 2004, 5).

In both New Zealand and Ireland the private sector (and indeed the public sector in Ireland) has been much more comfortable articulating a narrative of integrated solid waste management (ISWM) that is characterized by a strategy which utilizes a range of systems and processes to manage waste. Under this vision all methods of waste management, including resource recovery and landfill, are viable options provided attention is paid to environmental impacts. The key sub-narratives used under the ISWM schema are of efficiency, practicality and balance with the view that all waste management options, including recycling, impact on the environment and that no one single waste management option (including waste reduction) can alone offer a total solution given the diversity of waste being produced in modern society (Forfás 2003).

The relative influence of these narratives of waste management can be identified in the mechanisms that have been developed in New Zealand and Ireland. Ireland has introduced detailed legislation, the 1996 Waste Management Act (and amendments), that pays heed to the waste management hierarchy and calls for waste management planning regions to adopt an integrated waste management approach. In practice the major outcomes of this mechanism have been the development of recycling infrastructure across the country and plans for the development of municipal solid waste incinerators in particular locations. There has also been a consolidation of landfill sites and improvements in environmental protection measures installed. National targets for recycling of packaging and household were introduced in 1998 and landfill and plastic bag levies introduced nationally in 2001. More recently mechanisms have been introduced requiring local authorities to introduce pay-by-use waste charging systems. Although pay-by-use has not been made a statutory requirement for local authorities the majority have adopted some form of charging mechanism related to either weight or volume of waste collected. Research is underway to evaluate the impacts of this policy on volumes of waste collected and recycled as well as the knock-on effect it may have had on illegal waste practices such as fly-tipping (O'Callaghan-Platt and Davies 2006). In addition to market-based regulatory frameworks voluntary mechanisms, such as REPAK for packaging waste and the Race Against Waste awareness campaign for communities and small businesses, have been developed to assist in the achievement of recycling targets set out in national policy statements. In relation to these developments attention to waste prevention, the pinnacle of the waste hierarchy and the cornerstone of zero waste narratives, has been minimal with the development of a national waste prevention programme only emerging in 2004 with the first phase of demonstration projects funded in 2006.

Likewise in New Zealand there have been multiple developments in mechanisms to deal with waste with the 2002 National Waste Management Strategy at the forefront of policy interventions. This document, like the Waste Management Act in Ireland, called for more integration within the waste field although it did so from a much more explicit position with respect to waste minimization. Unlike Ireland there was no consideration of incineration as a means to reduce waste to landfill, although technically under the RMA a developer could propose an incinerator. While it has been said that this reiterates the zero waste approach to waste management, others have pointed out that there is no moratorium on incineration and provided the developer could prove there would be no adverse environmental effects it would be very difficult for a local authority to deny resource consents. The general consensus is that incineration is unlikely to emerge onto the New Zealand waste stream for a number of reasons. First, there exists a large and well co-ordinated opposition to incineration in powerful lobby groups such as farming as well as within environmental and community groups that any proposal would face huge costs in order to progress an application. Second, a number of large regional landfills have recently being consented thus providing capacity for much of projected waste volumes into the short to medium term and a high level of competition for any proposed incinerator. Third, there is no supra-national level of government imposing enforceable targets for diversion from landfill as there is in Ireland making the high capital investment costs of incineration economically unattractive.

Another area of divergence between the two countries lies in the use of market-based mechanisms for waste management. In New Zealand there has been no progression on a national waste levy and local authorities that have attempted to introduce levies within their jurisdiction have faced successful challenges in the courts. In line with its wider neo-liberal approach to governing current waste policy practice New Zealand clearly favours voluntary mechanisms for moving up the waste management hierarchy and attaining zero waste. For example producer responsibility schemes in NZ are voluntary and numerous, with schemes for packaging, electrical waste, tyres, waste oil, refrigerants and paint, while similar schemes in Ireland are fewer and some are legally binding (for example packaging, electrical waste and farm plastic). The aim of producer responsibility schemes is generally perceived as creating financial incentives to prevent or minimize end-of-life waste by redesigning products, but Forfás (2006) suggests that in Ireland at least the result has been for producers to finance the collective recycling of their product waste.

The final area of comparison in terms of outcomes relates to the nature and form of waste being produced. This would seem to be an obvious means through which to evaluate the effectiveness of different modes of governance operating in different locations. However such comparison is reliant on the existence of comparable, accurate and up-to-date information about the nature and volume of waste being collected and managed. In the 1990s it was recognized in both New Zealand and Ireland that information on waste was deficient. Both countries began a process of collating waste information and New Zealand produced its first national waste data report in 1997 with Ireland doing likewise in 1998. Ireland has produced reports on a regular basis ever since, but New Zealand has not conducted another national waste data collection exercise. Even in Ireland the quality of the information gathered for the national waste data reports is problematic for a variety of different approaches to data collection have been undertaken and the private sector interests have been reticent to reveal figures for fear of losing competitive advantage over rival companies. Unfortunately New Zealand does not use the municipal waste category when monitoring waste collected which also means that it is not possible to compare directly volumes of waste recycled or landfilled.

However a benchmarking report conducted by Forfás (2006) produced some comparisons based on communications with various Departments of Environment including those in Ireland and New Zealand. In terms of comparing facilities the benchmarking report established that New Zealand has 115 landfills compared to Irelands 35, however Ireland has around 30 biological treatment plants with less than 10 in New Zealand. In terms of recycling or reprocessing facilities New Zealand demonstrates more diversity and provision with three metal, four paper, one glass and one plastic facility whereas Ireland has only one glass and one plastic operation. Perhaps the starkest difference between the two countries is with respect to the ownership of municipal waste collection. New Zealand is dominated by sub-contracting to the private sector, with 80 per cent of collections operated this way, and the remainder evenly split between public ownership and purely private ownership. Ireland has a more even split between public and private ownership with only a small proportion (around 5 per cent) operated through sub-contracting. This means that for nearly 50 per cent of collected waste municipal authorities have no influence in the collection or disposal of waste leading to concerns over quality of waste management practices and pricing. Concerns over private sector practices in Ireland have led to discussions about developing a waste regulator to develop more consistency in standards for collections and recycling across the country. Although no such discussions are occurring in New Zealand there have been concerns about the dominance of a few key players in waste management collection and disposal activities.

Conclusion

This chapter has drawn together the experiences of waste management in Ireland and New Zealand that exhibit a number of common general characteristics but which articulate different discourses around waste issues and adopt contrasting positions in relation to certain waste management practices. More detailed analysis suggests that explanations for these divergent pathways, both discursive and material, are rooted in a complex intertwining of political, economic, social and environmental histories in the two countries. Examining the respective roles that tiers of government and spheres of governance play in the waste management of both countries further accounts for the existence of contrasting waste practices, but it is the distinct geographies of multi-level government that stand out as the most significant area of contrast between the two cases suggesting that while without doubt waste governance exists government still matters.

Chapter 8

Geographies of Garbage Governance: Some Concluding Thoughts

The concept of governance presents numerous advantages: it is flexible, adaptable, it takes nothing for granted, it encompasses a great diversity of actors and describes an ongoing process of interaction that is constantly changing in response to changing circumstances; it denotes a form of social coordination which can take into consideration various public and private interests in the management of matters of common concern and which takes responsibility for these matters collectively (Smouts 1998, 295).

The subject matter of politics is a buzzing, blooming confusion, unpredictable and violent. Our grasp of this world is fragile. All too often we simply seek to impose an order than is not there (Rhodes 1997, 200).

This book has presented a critical account of the role geography plays in environmental governance through the consideration of one significant environmental issue, municipal waste. Following Smouts (1998) a geographically informed analysis of governance has been established because, while previously perceived as primarily a technical issue for local governments, waste management now transcends localities and involves complex patterns of negotiated interactions that shape flows of waste production, treatment and disposal. These multiscalar and multiactor intersections ensure waste management is not simply a technical practice but also a highly political activity and as Rhodes (1997) suggests the world of politics is messy. So within all the confusion and complexity is it possible, or indeed appropriate, to try and create order from the chaos of waste governance? In part the answer to this question must be yes and in this final chapter a number of general points regarding the role of the state in managing waste, the problem of governance failure in waste governance and more theoretical issues related to governance and governmentality in the waste field are given further attention.

The State in Governance

It was proposed in Chapter 2 that one of the most significant issues for waste governance is examining the role and transformation of the state in managing waste given its historical position as the carrier of the collective interest (Pierre 2000). In the research conducted here, and supported by evidence from other waste studies detailed in Chapter 3, there does appear to be some evidence of a shift away from a linear (local) state-dominated political system to one that involves more complicated relations between various levels of state activity and non-state actors (Jessop 1994; Rosenau 1992). However rather than seeing these trends as a sign of the growing weakness of the nation-state it is possible to read them as strategies to renegotiate the power and authority of the state while devolving responsibility to other actors. There may be institutional transformations, for example through privatization of previously public waste services, which have restructured the participation of different actors in waste governing. In this research there are clear patterns of state transformation in relation to waste collection and disposal in both Ireland and New Zealand, albeit in different ways, that match with what Pierre (2000) calls a shift from a centripetal to a centrifugal form of governing. Here centripetal governing has the central state as the locus of political power and institutional capabilities, while centrifugal governing sees the state increasing its contact with other actors through processes or deregulation and decentralization. In New Zealand the local level of government retains the power to steer waste management and planning despite calls for more national government intervention and in spite of an increase in private sector and civil society provision of waste collection and disposal services. In Ireland local government also still has the statutory responsibility to ensure that waste is managed, but central government has been much more prescriptive in the ways that this occurs. This prescription has also involved considerable participation of private sector waste consultants in defining trajectories for waste management and a removal of waste management planning decisions from locally elected officials. In Ireland the limited use of sub-contracting within the privatization of the waste sector has raised concerns over the ability of the state at any scale to control the waste stream and this lack of authority has led to current discussions about the need for a waste regulator; a clear sign of the state seeking to reassert its authority over waste management.

However, in contrast to the findings of Stoker (2000) in relation to urban regimes and Rosenau (2000) who focuses on global interactions, there is less evidence that networks are a significant element in waste governance at least within the two case studies studied here. This is interesting as Chapter 2 identified networks as key players in the shift from government to governance as they transcend scales of government (from the local to the transnational) and incorporate a variety of actors from the public and private sectors as well as from civil society. It is not that networks or networking are absent from waste management, for example in both countries there are waste industry organizations, WasteMinz in New Zealand and the Irish Waste Management Association in Ireland. In addition there are transnational advocacy networks such as the Global Anti Incineration Alliance and Zero Waste Alliance that have a presence in both countries. New Zealand has more active national advocacy networks within the civil society sector with organizations such as Zero Waste New Zealand and the Community Recycling Network, while Ireland tends to have more localized and isolated waste related community campaigns or organizations. However industry and advocacy networks do not have an overt profile in waste management and planning fora. Instead these networks tend to work through informal personal contacts or in collaboration with broader organizations such as the Irish Business and Employers Confederation or the New Zealand Business Council for Sustainable Development in their attempts to influence policy trajectories. In part this may be attributed to the fairly recent emergence of waste network organizations. Zero Waste Alliance only initiated its campaign in Ireland in 2004 and the Global

Anti Incineration Alliance formed in 2000. While both WasteMinz and the Irish Waste Management Association have a longer lineage it is only in recent years that their membership rates have grown significantly. Yet while there is little evidence of cohesive waste networks within either the private or civil society sectors these organizations and their members can and do exert influence on policy discourses and could become more visible players in the future if networks become embedded, interactions between members become more dense and shifts towards the market and free enterprise within waste management continue.

Governance Failure

It was clearly identified in Chapters 5 and 6 that there are still significant challenges in relation to governing municipal waste in the countries studied and many of the challenges are similar to those identified in the overview of international waste management research presented in Chapter 3. Despite achievements in relation to increasing the amounts of waste recycled there has been little movement on waste minimization and prevention. Neither do consumption patterns seem to be changing radically and as a result volumes of waste continue to increase. More attention needs to be paid to why the current systems of governance have failed to achieve the end goals set out in the dominant discourses of the waste management hierarchy or integrated solid waste management that permeate both countries waste policies; that is to look more carefully at the causes and consequences of this failure of governance. One problem for local authorities is that privatized and contracted waste collection and disposal systems create fragmented networks of service providers who are more difficult to control, following Entwistle (1999, 376) while 'local authorities are principal players in these networks of governance ... their limited resources, flexibility and authority undermine their capacity to enable sustainable waste management'. There are, however, other reasons for governance failures that relate to the very conception and definition of waste and these are visible in the alternative discourses of resource stewardship and zero waste articulated in both countries. In the waste field these governance failures have led to conflicts over waste management. They are manifest in the alternative conceptions of governing waste, both in terms of calling for different practices and calling for the participation of different organizations. These points are clearly articulated by anti-incineration campaigners in Ireland and in the statements of opponents to waste facilities under the RMA in New Zealand.

Unsurprisingly, given the unpopularity of any requirements to modify behaviour, there is reluctance amongst elected officials in both Ireland and New Zealand to engage directly with the issue of consumption that leads to the production of waste. Yet it is with consumption practices that the whole waste cycle hinges and herein lies the contradiction. For while 'we carry Armageddon in our shopping bags' (Girling 2005, 2) waste is still identified as a badge of affluence so that 'pleasure, almost by definition, thrives on inessentials ... Conservationists, with their environmental audits and sustainability fetish, are the new puritans. They are the enemies of choice' (Girling 2005, 28). Given the reticence of governments to address conspicuous consumption, new technologies such as gasification and pyrolysis, which produce ash oils and synthetic gas to generate electricity and heat, and mechanical and biological treatments are being proposed as solutions to increasing volumes of waste. While potential advantages are heralded by these technologies in terms of reducing the volume of residual waste, sorting and separating materials and largely removing biodegradable elements they remain in competition with recycling processes and permit the continued consumption of resources. Therefore if the waste management hierarchy and integrated solid waste management are to retain their integrity the preoccupation with technical processes of disposing of waste will need to be superseded and the inherently political dimensions of consumption in relation to waste confronted.

Governance and governmentality

From a more theoretical perspective this research has confirmed that it is certainly still the case that we remain 'in a period of creative disorder concerning governance' (Kooiman, 2003, 5). However from this creative disorder in the governance literature it is possible to distil some basic theoretical models: the state-centric cascade (or trickledown or hierarchical) model; a system of co-governance where partnerships (of different types) between state and non-state actors at a variety of scales are formed; and finally a self-governance model based on a transfer of responsibility to non-state actors through privatization or deregulation (Symes 2006). However from the empirical findings of waste governance it seems that these models are not distinct categories but rather co-exist. For example there are elements of waste governance in Ireland that are clearly hierarchical, such as EU targets for diverting waste from landfill. Yet at the same time there are also partnerships between state and non-state actors through the drafting of waste planning at the county or regional level that suggest some degree of co-governance. Finally the large proportion of waste collection and disposal conducted by private waste companies without contracts with the public sector is indicative of self-government. In New Zealand the co-governance and self-government models seem to be more prevalent while the cascade model is less visible in the absence of a supranational governing structure. Nonetheless the local state, through the RMA and sub-contracting, still plays a significant role in steering the management of waste and, although not currently enacted, national government retains the power to impose more authority on the ways that waste management is organized.

Although the research indicated that attention to interactions between actors and organizations at different spatial scales is important it is also clear that there are different types of governance activity. These range from what Kooiman and Bavinck (2005) called the nitty-gritty of everyday actions, through broader institutional arrangements that include policy programmes to wider 'meta-governance' processes that involve broad values and general principles. Recognising these different types of governance activity has led some researchers, such as Braun (2000) and Drayton (2000), to turn their attention to the Foucauldian concept of governmentality in order investigate issues of power, authority and control that are central to patterns of environmental governance. To revisit briefly some of these ideas from Chapter 2, governmentality, as used by Foucault, implies an expansive way of thinking about governing and rule in relation to the exercise of modern power (Watts 2003, 9). A governmentality approach is less concerned with institutional arrangements and more interested in how the goals of governing are identified, usually defined as governmental rationalities, the mechanisms or governmental technologies which are used to try and achieve those goals and the relationship between those rationalities and technologies that leads to the formation of political authority. In sum the governmentality approach allows regimes of practice, or what Dean (1999, 17) calls 'organized ways of doing things', to be delineated.

Extending the work of Bulkeley et al. (2005) which highlights the dangers of underplaying the ability of individuals and networks to resist, contest or reinterpret these regimes of practice, this research has combined insights from across the governance-governmentality divide to inform a comparative study of waste, to examine the ways that waste is governed and the reasons why it is governed that way. Attention has been paid to both the multiple structures and processes of governing operating in various sites and through particular activities and the research has illuminated the existence of competing governmental rationalities, technologies and regimes of practice both between and amongst actors in the various spheres and tiers or waste governance.

Conclusion

On one level this study has shown that ordering the chaos that is waste governance by the application of a consistent method of data collection and common analytical framework is possible and appropriate. The geographies associated with waste management were analysed by examining how governments at different administrative scales interact with each other and with actors from public, private and civil society spheres of governance to create waste management policies and affect waste management outcomes; what might be termed the spatial signature of waste. Adopting this geographically sensitive governance analysis in an examination of two national contexts, Ireland and New Zealand, revealed the complexity and dynamism of processes within the waste governance arena. The comparison of results revealed both commonalities and differences in terms of governmental and governance structures, inter-actor relations and waste outcomes. But what do these findings say for wider attention to waste governance? The first point to make here is that the intention with this project was to create a framework for waste governance analysis that could be applied to different contexts and to different types of waste; to facilitate comparative waste governance research. However the research as conducted inevitably only provides a partial picture of the realities of governing waste and the limitations of the research can be delineated into two main areas. The first reflects the methodological difficulties of accommodating the micropolitics and multitude of everyday interactions between levels of government and between these levels of government and other spheres of governance. The second is related to the focus on municipal solid waste to the exclusion of other waste streams. While municipal solid waste is a particularly important because of its resonance with everyday practices and also by association a particularly challenging waste stream because of its diversity, other waste streams, such as industrial and agricultural by-products, are also significant contributors to overall waste production. It is quite likely that there are specific challenges resulting from these alternative sources, for example in terms of the toxicity of waste, which may lead to different patterns of governance. The test case provided here – examining municipal solid waste in Ireland and New Zealand – then presents simply the first step in what needs to be a broader consideration of waste governance. The framework will need to be tested more widely in diverse cultural, economic and political settings in order to build a more comprehensive picture of waste governance processes. In essence however, it is safe to conclude that there is governance of waste and that this governance is formed, informed and transformed by the actions and reactions of different tiers and spheres of governance that intersect in complex, asymmetric and dynamic ways. To concur with Elizabeth Royte's (2005) conclusion to her journey through garbage land, on the secret trail of trash, there is nothing more personal and local and nothing more inadvertently global than an individual's garbage.

Bibliography

Abrahamsen, R. (2000), *Disciplining Democracy: Development Discourse and Good Governance in Africa* (London: Zed Books).

Adeola, F. (2000), 'Cross-national Environmental Injustice and Human Rights Issues - A Review of Evidence in the Developing World', *American Behavioural Scientist* 43:4, 686-706.

Agarwal, A. and Narain, S. (1991), *Global Warming in an Unequal World* (New Delhi: Centre for Science and Environment).

Agarwal, A., Singhmar, A., Kulshrestha, M. and Mittal, A. (2005), 'Municipal Solid Waste Recycling and Associated Markets in Delhi, India', *Resources Conservation and Recycling* 44:1, 73-90.

Ahmed, S. and Ali, M. (2004), 'Partnerships for Solid Waste Management in Developing Countries: Linking Theories to Realities', *Habitat International* 28:3, 467-79.

Ali, M. (1999), 'The Informal Sector: What is it Worth?', *Waterlines*, 17:3, 10-2.

Allen, R. (2004), *No Global: the People of Ireland Versus the Multinationals* (Dublin: Pluto Press).

Anderson, A. (2002), 'A Fragile Plenty: Pre-European Māori and the New Zealand Environment', in Pawson, E. and Brooking, T. (eds.) *Environmental Histories of New Zealand* (Oxford: Oxford University Press). pp.19-34.

Auer, M. (2000), 'Who Participates in Global Environmental Governance? Partial Answers from International Relations Theory', *Policy Studies* 33:2, 155-80.

Aufrecht, S. (1999), 'Native American Governance in American Public Administration Literature', *American Review of Public Administration* 29:4, 370-90.

Awatere, S. (2003), *Tangata Whenua Perspectives* (Rotorua: New Zealand Land Collective).

Barlow, C. (1991), *Tikanga Whakaaro: Key Concepts in Maori Culture* (Auckland and New York: Oxford University Press).

Barr, S. (2002), *Household Waste in Social Perspective: Values, Attitudes, Situation and Behaviour* (Aldershot: Ashgate).

Beaumont, J. (2003), 'Governance and Popular Involvement in Local Authority Anti-poverty Strategies in the UK and the Netherlands', *Journal of Comparative Policy Analysis: Research and Practice* 5:2-3, 189-207.

Beck, U. (1995), *Ecological Politics in an Age of Risk* (Cambridge: Polity Press).

Beder, S. and Shortland, M. (1992), 'Siting a Hazardous Waste Facility: The Tangled Web of Risk Communication', *Public Understanding of Science* 1:2, 139-60.

Begum, R. (1999), 'Is it the Responsibility of the Poor Children to Clean our Cities?', *Chinta* 8:1, 35.

Bell, C. (1996), *Inventing New Zealand: Everyday Myths of Pakeha Identity* (Auckland: Penguin Books).

Berke, P., Crawford, J., Dixon, J. and Ericksen, N. (1999), 'Do Co-operative Environmental Planning Mandates Produce Good Plans? Empirical Results from the New Zealand Experience', *Environment and Planning B: Planning and Design* 26:5, 643-64.

Berke, P., Ericksen, N., Crawford, J. and Dixon, J. (2002), 'Planning and Indigenous People: Human Rights and Environmental Protection in New Zealand', *Journal of Planning Education and Research* 22:2, 115-34.

Berke, P., Backhurst, M., Day, M., Ericksen, N., Laurian, L., Crawford, J. and Dixon, J. (2006), 'What Makes Plan Implementation Successful? An Evaluation of Local Plans and Implementation Practices in New Zealand', *Environment and Planning B: Planning and Design* 33:4, 581-600.

Bevan, J. and Jay, M. (1998), 'Compensation for Adverse Effects, or Buying Consent?', *Resource Management News* 6:1, 4-7.

Blaikie, P. and Brookfield, H. (1987), *Land Degradations and Society* (London: Methuen).

Blowers, A. (1998), 'Power, Participation and Partnership: the Limits of Co-operative Environmental Management', in Glasbergen, P. (ed.) *Co-operative Environmental Management Agreements as a Policy Strategy* (Dordrecht, London: Kluwer). pp.229-50.

Boyle, M. (2001), 'Cleaning up After the Celtic Tiger: the Politics of Waste Management in the Irish Republic', *Journal of Scottish Association of Geography Teachers* 30:1, 71-91.

_____ (2003), 'Scale as an Active Progenitor in the Metamorphosis of the Waste Management Hierarchy in Member States. The Case of the Republic of Ireland', *European Planning Studies* 11:4, 481-502.

Boyle, C. (2000), 'Solid Waste Management in New Zealand', *Waste Management* 20:7, 517-26.

Bradshaw, K. (2004), *Reduce Your Rubbish: Lessons from this National Pilot Campaign* (Wellington: Ministry for the Environment).

Braun, B. (2000), 'Producing Vertical Territory', *Ecumene* 7:1, 7-46.

Britton, S., Le Heron, R., and Pawson, E. (1992), *Changing Places in New Zealand: A Geography of Restructuring* (Christchurch: New Zealand Geographical Society).

Brodnax, R. and Milne, J. (2002), 'Towards Zero Waste and a Sustainable New Zealand', *Journal of the Resource Management Law Association* 10:2, 23-7.

Brown, G. (2000), *You Too Can Profit from Cleaner Production and Waste Minimisation, Brochure Reporting the Results of the Target Zero Project* (Wellington: Ecosense Ltd).

Bryant, R. and Bailey, S. (1997), *Third World Political Ecology* (London: Routledge).

Bührs, T. (2003), 'From Diffusion to Defusion: The Roots and Effects of Environmental Innovation in New Zealand', *Environmental Politics* 12:3, 83-101.

Bührs, T. and Bartlett, R. (1993), *Environmental Policy in New Zealand: The Politics of Clean and Green?* (Auckland: Oxford University Press).

Bulkeley, H. (2000), 'Discourse Coalitions and the Australian Climate Change Policy Network', *Environment and Planning C: Government and Policy* 18, 727-748.

_____ (2005), 'Reconfiguring Environmental Governance: Towards a Politics of Scales and Networks, *Political Geography* 24:8, 875-902.

Bulkeley, H. and Betsill, M. (2003) *Cities and Climate Change: Urban Sustainability and Global Environmental Governance* (London: Routledge).

Bulkeley, H., Davies, A., Evans, B., Gibbs, D., Kern, K. and Theobald, K. (2003), 'Environmental Governance and Transnational Local Authority Networks in Europe', *Journal of Environmental Politics and Planning* 5:3, 235-54.

Bulkeley, H., Watson, M., Hudson, R. and Weaver, P. (2005), 'Governing Municipal Waste: Towards a New Analytical Framework', *Journal of Environmental Policy and Planning* 7:1, 1-23.

_____ (2007), 'Modes of Governing Municipal Waste', *Environment and Planning A* 39:11, 2733-2753.

Bullard, R. (2000), *Dumping in Dixie: Race, Class and Environmental Quality* (Boulder CO: Westview Press).

Burnley, S., Ellis, J., Flowerdew, R., Poll, A. and Prosser, H. (2007), 'Assessing the Composition of Municipal Solid Waste in Wales', *Resources, Conservation and Recycling* 49:3, 264-83.

Cabot, D. (1999), *Ireland: A Natural History* (London: Harper Collins).

CAE (1992), *Our Waste: Our Responsibility* (Christchurch: Centre for Advanced Engineering).

Callanan, M. and Keogan, J. (2003), *Local Government in Ireland: Inside Out* (Dublin: Institute of Public Administration).

Camacho, D. (1999), *Environmental Injustices, Political Struggles: Race, Class and the Environment* (Durham: Duke University Press).

Cameron, L. (2002), *Changing Waste Minimisation and Other Environmental Behaviours through Community Interventions* (Hamilton: Environment Waikato).

Carnoy, M. and Castells, M. (2001), 'Globalisation, the Knowledge Society and the Network State', *Global Networks* 1:1, 1-19.

Carolan, J. (2000), *Save the Waikato River – Dump the Dump* (Auckland: Socialist Workers Organisation).

Carolan, M. (2004), 'Ecological Modernisation Theory: What About Consumption?', *Society and Natural Resources* 17:3, 247-60.

Castells, M. (1997), *The Power of Identity* (Oxford: Blackwell).

CEC (Commission of the European Communities) (2001), *European Governance, A White Paper*, COM (2001) 428 Final (CEC, Brussels).

Chalfan, B. (2001), *Zero Waste: Key to Our Future. The Case for Zero Waste* (Portland OR: Zero Waste Alliance).

Chapple, G. (1995), 'Clean, Green and Expensive' *NZ Listener*, 22 July, pp.18-20.

Chaturvedi, B. (1998), *Public Waste, Private Enterprise* (Berlin: Heinrich Boel Stiftung).

Clarke, M., Read, A. and Phillips, P. (1999), 'Integrated Waste Management Planning and Decision Making in New York City', *Resources Conservation and Recycling*, 26:2, 125-41.

Cocklin, C. and Furuseth, O. (1994), 'Geographical Dimensions of Environmental Restructuring in New Zealand' *Professional Geographer* 46:4, 459-67.

Comhar (2002), *Principles of Sustainable Development* (Dublin: Comhar).

Commerce Commission (1999), *Decision 355 Clearance of Business Acquisition* (Wellington, Commerce Commission).

CONAMA (2006), *Draft National Strategy for Prevention and Minimization of Solid Wastes* (Santiago: Chile National Environment Commission).

Connett, P. and Sheehan, B. (2001), *A Citizen's Guide for Zero Waste: A US and Canadian Perspective* (Cotati CA: Grassroots Recycling Network).

Connolly, J. (1997), *Beyond the Politics of Law and Order: Towards Community Policing in Ireland* (Belfast: Belfast Centre for Research and Documentation).

Council of Europe (2005), *Nature, Culture and Landscape for Sustainable Spatial Development* (Strasbourg: Directorate of Culture and Culture and Natural Heritage).

Cox, K. (1998), 'Spaces of Dependence, Spaces of Engagement and the Politics of Scale, or: Looking for Local Politics' *Political Geography* 17:1, 1-23.

Coyle, F. and Fairweather, J. (2005), 'Challenging a Place Myth: New Zealand's Clean Green Image Meets the Biotechnology Revolution', *Area*, 37:2, 148-158.

Crittenden, B. and Kolaczkowski, S. (1995), *Waste Minimisation: a Practical Guide* (Rugby: Institution of Chemical Engineers).

CSO (2006), *2006 Census: Preliminary Report* (Dublin: The Stationery Office).

Cunningham, J. and Clinch, P. (2005), 'Innovation and Environmental Voluntary Approaches', *Journal of Environmental Planning and Management* 48:2, 373-92.

Dann, C. (2002), 'Losing Ground? Environmental Problems and Prospects at the Beginning of the Twenty-First Century', in Pawson E. and Brooking, T. (eds.) *Environmental Histories of New Zealand* (Oxford: Oxford University Press).

Darier, E. (1999), *Discourses of the Environment* (Oxford: Blackwell Publishers).

Davidson, D. and Frickel, S. (2004), 'Understanding Environmental Governance', *Organization and Environment*, 17:4, 471-92.

Davies, A. (2002), 'Power, Politics and Networks: Shaping Partnerships for Sustainable Communities', *Area* 34:2, 190-203.

_____ (2003), 'Waste Wars – Public Attitudes and the Politics of Place in Waste Management Strategies' *Irish Geography* 36:1, 77-92.

_____ (2004), 'Working with Waste', *Waste Awareness*, January/February 2004, 6.

_____ (2005), 'Incineration Politics and the Geographies of Waste Governance: A Burning Issue For Ireland?', *Environment and Planning C: Government and Policy* 23:3, 373-298.

_____ (2006), 'Anti-incineration Campaigning in Ireland: A Case for Developing an Environmental Justice Dialogue?', *Geoforum* 37:5, 708-24.

_____ (2007), 'A Wasted Opportunity? Civil Society and Waste Management in Ireland', *Environmental Politics* 16:1, 52-72.

_____ (forthcoming) 'Civil Society Activism and Waste Management in Ireland: The Carranstown Anti-incineration Campaign', *Land Use Policy.*

Davies, A., Fahy, F. and Taylor, D. (2005), 'Mind the Gap! Householder Attitudes and Actions Towards Waste in Ireland', *Irish Geography* 38:2, 151-68.

Davoudi, S. (2000), 'Planning for Waste Management: Changing Discourses and Institutional Relationships' *Progress in Planning* 53:3, 165-216.

Davoudi, S. and Evans, N. (2005), 'The Challenge of Governance in Regional Waste Planning', *Environment and Planning C: Government and Policy* 23:4, 493-517.

Day, M., Backhurst, M., Erickson, N., Crawford, J., Chapman, S., Berke, P., Laurian, L., Dixon, J., Jefferies R., Warren, T., Barfoot, C., Mason, G., Bennett, M. and Gibson, C. (2003), *District Plan Implementation under the RMA: Confessions of a Resource Consent* (Hamilton: University of Waikato).

de Graf, J., Wann, D. and Naylor, T. (2002), *Affluenza: The All Consuming Epidemic* (San Francisco: Barrett-Koelher).

Dean, M. (1999), *Governmentality: Power and Rule in Modern Society* (London: Sage).

DEFRA (2004), Key Facts about Waste and Recycling (London: DEFRA). <http://www.defra.gov.uk/environment/statistics/waste/kf/wrk08.htm>, accessed 21 July 2006.

DeLillo, D. (1997), *Underworld* (New York: Scribner).

Dente, B., Fareri, P. and Ligteringen, J. (1998), *The Waste and the Backyard: The Creation of Waste Facilities: Success Stories in Six European Countries* (Dordrecht: Kluwer Academic Press).

Department of Environmental Affairs and Tourism (1999), *National Waste Management Strategy: National Waste Management Strategies and Action Plans, South Africa*, Version D, PMG 130, PSC 69 (Cape Town: Department of Environmental Affairs and Tourism).

Department of Finance (2006), *Monthly Economic Bulletin, November 2006* (Dublin: Department of Finance).

Department of Prime Minister and Cabinet (2003), *Sustainable Development for New Zealand: Programme of Action* (Wellington: DPMC).

Department of Taoseach (2005), *Sustaining Progress: 2003-2005 Final Report on Special Initiatives* (Dublin: Stationery Office).

Dew, K. (1999), 'National Identity and Controversy: New Zealand's Clean Green Image and Pentachlorophenol', *Health and Place* 5, 45-57.

Di Chiro, G. (1997), 'Local Actions, Global Visions: Remaking Environmental Expertise', *Frontiers: A Journal of Women Studies* 8, 203–31.

Dirlik, A. (1999), 'Place-based Imagination: Globalism and the Politics of Place', *Review: A Journal of the Fernand Braudel Center* 22:2, 151-87.

DoEHLG (2003), *Protection of the Environment Act* (Dublin: DoEHLG - Department of the Environment, Heritage and Local Government).

_____ (2004), *Interreg IIIA: Cross Border Awareness Campaign: Summary Report and Evaluation* (Dublin: DoEHLG - Department of the Environment, Heritage and Local Government).

_____ (2006a), *Roche Provides €4m for Local Authorities to Support the Operating Costs of Recycling Facilities* (Dublin: DoEHLG - Department of the Environment, Heritage and Local Government).

_____ (2006b), *Regulation of the Waste Management Sector: Consultation Paper* (Dublin: DoEHLG - Department of the Environment, Heritage and Local Government).

DoELG (1996), *1996 Waste Management Act* (Dublin: DoELG – Department of the Environment and Local Government).

_____ (1997), *Sustainable Development: A Strategy for Ireland* (Dublin: DoELG – Department of the Environment and Local Government).

_____ (1998a), *An Overview of the Waste Management Act 1996* (Dublin: DoELG – Department of the Environment and Local Government).

_____ (1998b), *Changing Our Ways: Policy Statement* (Dublin: DoELG – Department of the Environment and Local Government).

_____ (2000), *National Development Plan* (Dublin: DoELG – Department of the Environment and Local Government).

_____ (2002), *Preventing and Recycling Waste: Delivering Change* (Dublin: DoELG – Department of the Environment and Local Government).

Donahue, J. (2002), 'Market-based Governance and the Architecture of Accountability', in Donahue, J. and Nye, J. (eds) *Market-Based Governance* (Washington: Brookings Institution). pp.201-26.

Dong, S., Tong, K. and Wu, Y. (2001), 'Municipal Solid Waste Management in China: Using Commercial Management to Solve a Growing Problem', *Utilities Policy* 10, 7-11.

Donnelly, J. (2000), *The Great Irish Potato Famine* (London: Sutton Publishing).

Dooney, S. and O'Toole, J. (1998), *Irish Government Today* (Dublin: Gill and Macmillan).

Drayton, R. (2000), *Nature's Government: Science, Improvement and Imperial Britain and the Improvement of the World* (New Haven: Yale University Press).

Dunlap, T. (1999), *Nature and the English Diaspora* (Cambridge: Cambridge University Press).

Eberlein, B. and Kerwer, D. (2004), 'New Governance in the EU: A Theoretical Perspective', *Journal of Common Market Studies* 42:1, 121-42.

EC (1994), *European Parliament and Council Directive on Packaging and Packaging Waste 92/62/EC* (Brussels: EC - European Council).

Eckstein, H. (1975), 'Case Study Theory in Political Science', in Greenstein, F. and Polsby, N. (eds), *Handbook of Political Science Vol. 7 Strategies of Inquiry,* (Reading MA: Addison-Wesley).

Eggerth, L. (2005), 'The Evolving Face of Private Sector Participation in Solid Waste Management', *Waste Management* 25:3, 229-30.

El-Fadel, M. and Khoury, R. (2001), *Municipal Solid Waste Management in Lebanon: Impact Assessment, Mitigation, and the Need for an Integrated Approach. Technical Report, FEA-CEE-2001-01* (Lebanon: United States Agency for International Development (USAID), American University of Beirut).

Entwistle, T. (1999), 'Towards Sustainable Waste Management: Central Steering, Local Enabling or Autopoiesis?', *Policy and Politics* 27:3, 375-88.

Environment Court (2001), *Hampton Downs Environment Court Decision No. A 110/01* (Auckland: Environment Court).

Envision (2003a), *Getting There: The Road to Zero Waste. Strategies for Sustainable Communities* (Takapuna: Envision and ZWNZ).

_____ (2003b), *Resourceful Communities: A Guide to Resource Recovery Centres in New Zealand* (Auckland: Envision New Zealand).

EPA (2000), *National Waste Database Report 1998* (Wexford: EPA - Environmental Protection Agency).

_____ (2001), *Municipal Solid Waste in the US, 2000 Facts and Figures* (Washington: EPA – Environmental Protection Agency).

_____ (2004), *Waste Management – Taking Stock and Moving Forward* (Wexford: EPA – Environmental Protection Agency).

_____ (2005), *National Waste Report 2004* (Wexford: EPA – Environmental Protection Agency).

Erickson, N., Berke, P., Crawford, J. and Dixon, J. (2003), *Planning for Sustainability: New Zealand under the RMA* (Aldershot: Ashgate).

Ericksen, N., Berke, P. and Crawford, J. (2004), *Planning for Sustainability: The New Zealand Experience* (Aldershot, Hants: Ashgate).

Escobar, A. (1996), 'Constructing Nature: Elements for a Post-structuralist Political Ecology', in Peet, R. and Watts, W. (eds). *Liberation Ecologies: Environment, Development and Social Movements* (London: Routledge). pp.46-68.

Fagan, H, (2004), 'Waste Management and its Contestation in the Republic of Ireland', *Capitalism Nature Socialism* 15:1, 83-102.

Fairbrass, J. (2003), *Environmental Governance and the Dispersal of Decision-making Working paper EDM 03-10* (Norwich: CSERGE).

Faughnan, P. and McCabe, B. (1998), *Irish Citizens and the Environment: A Cross-National Study of Environmental Attitudes, Perceptions and Behaviours* (Dublin: Environmental Protection Agency).

Featherstone, D. (2003), 'Spatialities of Transnational Resistance to Globalization: The Maps of Grievance of the Inter-Continental Caravan', *Transactions of the Institute of British Geographers NS* 28:4, 404–21.

Fischel, W. (2001), 'Why are there NIMBYs?' *Land Economics* 77:1, 142-52.

Fleming, G. (2002), *Clean, Green Image Under Threat* (Wellington: The Dominion Post).

Foley, M. and Edwards, B. (1996), 'The Paradox of Civil Society', *Journal of Democracy* 7:3, 38-52.

Forfás (2001), *Key Waste Management Issues in Ireland* (Dublin: Forfás).

_____ (2006), *Waste Management Benchmarking Study: A Baseline Assessment* (Dublin: Forfás).

Forsyth, T. (2005), 'Building Deliberative Public-Private Partnerships for Waste Management in Asia', *Geoforum* 36:4, 429-39.

Foster, J. (1997), *Nature in Ireland: A Scientific and Cultural History* (Dublin: Lilliput Press).

Frame, B. (2004), 'The Big Clean Up: Social Marketing for the Auckland Region', *Local Environment* 9:6, 507-26.

Gandy, M. (2001), *Recycling and the Politics of Urban Waste* (Basingstoke: Palgrave Macmillan).

_____ (2002), *Concrete and Clay: Reworking Nature in New York City* (Cambridge, Massachusetts: MIT Press).

Gardner, D. (2003), *Measuring the National Reduce Your Rubbish Campaign* (Wellington: MfE – Ministry for the Environment).

Garner, R. (2000), *Environmental Politics: Britain, Europe and the Global Environment* (Basingstoke: Macmillan Press).

Gertsakis, J. and Lewis, H. (2003), *Sustainability and the Waste Management Hierarchy: A Discussion Paper for Ecorecycle* (Melbourne: Victoria, RMIT).

Girling, R. (2005), *Rubbish! A Chronicle of Waste* (London: Eden Books, Transworld).

Glasbergen, P. (1998), *Co-operative Environmental Management Agreements as a Policy Strategy* (London: Kluwer, Dordrecht).

González, S. and Healey, P. (2005), 'A Sociological Institutionalist Approach to the Study of Innovation in Governance Capacity', *Urban Studies* 42:11, 2055-69.

Gordon, C. (1991), 'Governmental Rationality: An Introduction', in Burchell, G., Gordon, C., and Miller, P. (eds) *The Foucault Effect: Studies in Governmentality* (Chicago: University of Chicago Press). pp.1-51.

Gray, P. (1995), 'Waste Incineration – Controversy and Risk Communication', *European Review of Applied Psychology* 45:1, 29-34.

Green Alliance (2002), *Creative Policy Packages for Waste: Lessons for the UK* (London: Green Alliance).

Green Party (2004), *Sargent Accuses Cullen of a 'Bury and Burn' Policy* (Dublin: Green Party).

Greenpeace (1993), *The International Trade in Toxic Wastes: An International Inventory* (Washington: Greenpeace International).

Grossman, P. (2000), 'Corporate Interest and Trade Liberalization: The North American Free Trade Agreement and Environmental Protection' *Organization and Environment* 13:1, 61-85.

Grundy, K. and Gleeson, B. (1996), 'Sustainable Management and the Market: The Politics of Planning Reform in New Zealand', *Land Use Policy* 13:3, 197-211.

Gunder, M., and Mouat, C. (2002), 'Symbolic Violence and Victimization in Planning Processes: A Reconnoitre of the New Zealand Resource Management Act', *Planning Theory* 1:2, 124-45.

Hajer, M. (1995), *The Politics of Environmental Discourse: Ecological Modernization and the Policy Process* (Oxford: Oxford University Press).

_____ (2003), 'Policy Without Polity? Policy Analysis and the Institutional Void', *Policy Sciences* 36:2, 175-95.

Halfini, M. (1997), 'The Challenge of Urban Governance in Africa: Institutional Change and the Knowledge Gap', in Swilling, M. (ed) *Governing Africa's Cities*, (Johannesburg: Witwatersrand University Press). pp.13-34.

Halla, F. and Majani, B. (1999), 'Innovative Ways for Solid Waste Management in Dar-es-Salaam: Toward Stakeholder Partnerships', *Habitat International* 23:3, 351-61.

Hansen, K. and Vaa, M. (2004), *Reconsidering Informality: Perspectives from Urban Africa* (Uppsala, Sweden: Nordic Africa Institute).

Hanson, S. and Lake, R. (2000), 'Needed: Geographic Research on Urban Sustainability', *Urban Geography* 21:1, 1-4.

Hardoy, J., Mitline, D. and Satterthwaite, D. (1992), *Environmental Problems in Third World Cities* (London: Earthscan).

Harris, P. and Twiname, L. (1998), *First Knights: An Investigation of the New Zealand Roundtable* (Auckland: Howling at the Moon Publishing).

Healey, P. (1997), *Collaborative Planning: Shaping Places in Fragmented Societies* (Basingstoke, Macmillan).

Hearn, T. (2002), 'Mining the Quarry', in Pawson, E. and Brooking, T. (eds), *Environmental Histories of New Zealand* (Oxford: Oxford University Press). pp.84-99.

Henry, R., Yongsheng, Z. and Jun., D. (2006), 'Municipal Solid Waste Management Challenges in Developing Countries – Kenyan Case Study', *Waste Management* 26:1, 92-100.

Herbert-Cheshire, L. (2003), 'Translating Policy: Power and Action in Australia's Country Towns', *Sociologia Ruralis* 43:4, 454-73.

Higgens, M. (1982), 'The Limits of Clientilism: Towards an Assessment of Irish Politics', in Clapham, C. (ed.), *Private Patronage and Public Power* (London: Frances Pinter).

Hill, J., Bégin, A., and Shaw, B. (2002), *Creative Policy Packages for Waste: Lessons for the UK* (London: Green Alliance).

Hirst, P. (2000), 'Democracy and Governance', in Pierre, J. (ed.), *Debating Governance* (Oxford: Oxford University Press). pp.13-25.

Hooghe, L. and Marks, G. (1997), 'Contending Models of Governance in the European Union', in Cafruny, A. and Lanowski, C. (eds), *Europe's Ambiguous Unity: Conflict and Consensus in the Post-Maastricht Era* (Boulder and London: Lynne Rienner Publishers). pp.21-33.

_____ (2003), 'Unravelling the Central State, But How?', *American Political Science Review* 97:2, 233-43.

Howe, S. (2000), *Ireland and Empire: Colonial Legacies in Irish History and Culture* (Oxford: Oxford University Press).

Hughey, K., Kerr, G. and Cullen, R. (2004), *Perceptions of the State of the Environment: the 2004 (3rd Biennial) Survey of Public Attitudes, Preferences and Perceptions of the New Zealand Environment* (Lincoln: Environmental Management Group, Lincoln University).

Hui, Y., Li'ao, W., Fenwei, S. and Gang, H. (2006) 'Urban Solid Waste Management in Chongqing: Challenges and Opportunities', *Waste Management* 26:9,1052-62.

Hunter, S. and Leyden, K. (1995), 'Beyond NIMBY: Explaining Opposition to Hazardous Waste Facilities', *Policy Studies Journal* 23:4, 601-19.

Hyden, G. (1999), 'Governance and the Reconstitution of Political Order', in Joseph, R. (ed), State, Conflict and Democracy in Africa (Boulder and London: Lynne Reiner Publishers). pp.179-96.

Indecon (2005), *Indecon Review of Local Government Financing* (Birmingham: Indecon Economic Consultants with Institute of Local Government Studies, University of Birmingham, Birmingham).

Irish Examiner (2001), 'The Tricky Business of Making the Emerald Isle Go Green', *Irish Examiner*, 1st February 2001, p.10.

Itibayo, O. (2002), 'Public-Private Partnerships in the Siting of Hazardous Waste Facilities: the Importance of Trust', *Waste Management and Research* 20:3, 212-22.

Jackson, T. and Dixon, J. (2004), 'The New Zealand Resource Management Act: an Exercise in Delivering Sustainable Development through an Ecological Modernisation Agenda', *Planning & Development Research Conference: People and Place - Common Problems, Shared Solutions* (Aberdeen, University of Aberdeen) 31 March-2 April 2004.

Jessop, B. (1994), 'Post-Fordism and the State', in A. Amin (ed.), *Post-Fordism: A Reader* (Oxford: Blackwell). pp.251-79.

_____ (1995), 'The Regulation Approach, Governance and Post-Fordism: Alternative Perspectives on Economic and Political Change', *Economy and Society* 24:3, 307-33.

_____ (1999), 'The Changing Governance of Welfare: Recent Trends in its Primary Functions, Scale, and Modes of Co-ordination', *Social Policy and Administration* 33:4, 348-59.

_____ (2004), 'The Political Economy of Scale and European Governance', *Tijdschrift voor Economische en Sociale Geografie* 96:2, 225-30.

Jordan, A. (2001), 'The European Union: an Evolving System of Multi-Level Governance...or Government?', *Policy and Politics* 29:2, 193-208.

Jordan, A. and Schout, A. (2005), 'Coordinated European Governance: Self-Organising or Centrally-Steered?', *Public Administration* 83:1, 201-20.

Jordan, A., Wurzel, R. and Zito, A. (2005), 'The Rise of 'New' Policy Instruments in Comparative Perspective: Has Governance Eclipsed Government?', *Political Studies*, 53:4, 477-96.

Kaseva, M. and Mbuligwe, S. (2005), 'Appraisal of Solid Waste Collection in Dar es Salaam City, Tanzania', *Habitat International* 29:2, 353-66.

Keck, M. and Sikkink, K. (1998), *Activists Beyond Borders: Advocacy Networks in International Politics* (Ithaca and London: Cornell University Press).

Kelly, M., Kennedy, F., Faughnan, P. and Tovey, H. (2003a), *Cultural Sources of Support on which Environmental Attitudes and Behaviours Draw: Second Report of National Survey Data* (Dublin: EPA – Environmental Protection Agency).

_____ (2003b), *Environmental Attitudes and Behaviours: Ireland in Comparative European Perspective: Third Report of National Survey Data* (Dublin: EPA – Environmental Protection Agency).

Kelly, O. (2006), 'Bord Receives over 2000 Incinerator Objections', *Irish Times*, 3/10/06, p.7.

Kelsey, J. (1997), *The New Zealand Experiment: A World Model for Structural Change?* (Auckland, Auckland University Press).

Keynote (2007), *Global Waste Management* (Hampton: Keynote).

Kiel, K. and McClain, K. (1996), 'House Price Recovery and Stigma after a Failed Siting', *Applied Economics* 28:11, 1351-8.

Kironde, J. (2001), *Urban Poverty in Tanzania: The Role of Urban Authorities* (Dar es Salaam: University College of Lands and Architectural Studies).

Kironde, J. and Yhdego, M. (1997), 'The Governance of Waste Management in Urban Tanzania: Towards a Community Based Approach', *Resources, Conservation and Recycling* 21:4, 213-26.

Kironde, J. and Ngware, S. (2000), Introduction, in Ngware, S. and Kironde, J. (eds), *Urbanising Tanzania: Issues, Initiatives and Priorities* (Dar es Salaam: University of Dar es Salaam Press). pp.1-6.

Kjaer, A. (2004), *Governance* (Oxford: Polity).

Klooster, D. (2005), 'Environmental Certification of Forests: The Evolution of Environmental Governance in a Commodity Network', *Journal of Rural Studies* 21:4, 403-17.

Kolk, A. (2000), *Economics of Environmental Management* (London: Financial Times-Prentice Hall).

Kooiman, J. (2003), *Governing as Governance* (London: Sage).

Kooiman, J. and Bavinck, M. (2005), 'The Governing Perspective', in: Kooiman, J., Jentoft, S., Pullin, R., and Bavinick, M. (eds.), *Fish for Life: Interactive Governance for Fisheries* (Amsterdam: Amsterdam University Press). pp.11–24.

Kubal, T. (1998), 'The Presentation of Political Self: Cultural Resonance and the Construction of Collective Action Frames', *Sociological Quarterly* 39:4, 539-554.

Kuhn, R., and Bullard, K. (1998), 'Canadian Innovation in Siting Hazardous Waste Management Facilities', *Environmental Management* 22:4, 533-45.

Kutting, G. (2000), *Environment, Society and International Relations: Towards More Effective International Environmental Agreements* (London: Routledge).

Laffen, B. (1996), 'Ireland: a Region within Regions – the Odd Man Out', in Hooge, L. (ed.), *Cohesion Policy and European Integration: Building Multi-level Governance* (Oxford: Oxford University Press). pp.320-37.

Lane, M. (2006), 'The Role of Planning in Achieving Indigenous Land Justice and Community Goals', *Land Use Policy* 23:4, 385-94.

Larragy, J. (2006), 'Origins and significance of the Community-Voluntary Pillars Entry to Irish Social Partnership', *Economic and Social Review* (Dublin: ESRI).

Le Heron, R. and Pawson, E. (1996), *Changing Places: New Zealand in the 1990s* (Auckland: Longman).

Leach, H. (2002), 'Exotic Natives and Contrived Wild Gardens: The Twentieth Century Home Garden', in Pawson, E. and Brooking, T. (eds.). *Environmental Histories of New Zealand* (Oxford: Oxford University Press). pp.214-32.

Leach, R. and Percy-Smith, J. (2001), *Local Governance in Britain* (Basingstoke: Palgrave).

Leonard, H. (1988), *Pollution and the Struggle for the World Product: Multinational Corporations, Environment and International Comparative Advantage* (Cambridge: Cambridge University Press).

Leonard, L. (2006), *Green Nation: The Irish Environmental Movements from Carnsore Point to the Rossport Five* (Drogheda: Choice Publishing).

Liftin, K. (1994), *Ozone Discourses: Science and Politics in Global Environmental Cooperation* (New York: Columbia University Press).

Lima, M. (2004), 'On the Influence of Risk Perception on Mental Health: Living Near an Incinerator', *Journal of Environmental Psychology* 24:1, 71-84.

Lipschutz, R. (1997), 'Networks of Knowledge and Practice: Global Civil Society and Protection of the Global Environment', in Brooks, L. and VanDeveer, S.

(eds.), *Saving the Seas: Values, Scientists and International Governance* (College Park, MD: Maryland Sea Grant College). pp.427-68.

Louis, E. (2004), 'A Historical Context of Municipal Waste Management in the United States', *Waste Management and Research* 22:4, 306-22.

Luckin, D. and Sharp, L. (2004), 'Remaking Local Governance through Community Participation? The Case of the UK Community Waste Sector', *Urban Studies* 41:8, 1485-505.

Luloff, A., Albrecht, S. and Bourke, L. (1998), 'NIMBY and the Hazardous and Toxic Waste Siting Dilemma: The Need for Concept Clarification', *Society and Natural Resources* 11:1, 81-9.

Lund-Thomsen, P. (2005), 'Corporate Accountability in South Africa: The Role of Community Mobilizing in Environmental Governance', *International Affairs*, 81(3): 619-34.

MacLeod, G. and Goodwin, M. (1999), 'Space, Scale and State Strategy: Rethinking Urban and Regional Governance', *Progress in Human Geography* 23:4, 503-27.

Majani, B. (2002), 'Environmental Planning and Management in Dar es Salaam: A Global Success Story and Learning Experience', *Journal of Building and Land Development* 9:1, 67-72.

Marinetto, M. (2003), 'Governing Beyond the Centre', *Political Studies* 51:3, 592-608.

Marsh, D. and Rhodes, R. (1992), *Policy Networks in British Government* (Oxford: Clarendon Press).

Martello, M. and Jasanoff, S. (2004), 'Introduction: Globalization and Environmental Governance', in Martello, M. and Jasanoff, S. (eds.) *Earthly Politics.* (Cambridge MA: MIT Press). pp.1-30.

Massey University (2001), *New Zealanders and the Environment* (Palmerston North: Massey University, Department of Marketing).

Massoud, M. and El-Fadel, M. (2002), 'Public-Private Partnerships for Solid Waste Management Services', *Environmental Management* 30:5, 621-30.

McAleese, D. (2000), 'The Celtic Tiger Origins and Prospects', *Policy Options* July August, 2000, 46-50.

McAvoy, G. (1999), *Controlling Technocracy: Citizen Rationality and the NIMBY Syndrome* (Georgetown: Georgetown University Press).

McCarthy, J. (2005), 'Scale, Sovereignty and Strategy in Environmental Governance', *Antipode*, 37:4, 731-53.

McDonald, B. (2006), *An Introduction to Sociology in Ireland* (Dublin: Gill Macmillan).

McDonald, F. and Nix, J. (2005), *Chaos at the Crossroads* (Cork: Gandon).

McDougall, F., White, P., Franke, M. and Hindel, P. (2001), *Integrated Solid Waste Management: A Life Cycle Inventory* (2nd Ed) (Oxford: Blackwell Science).

McKenzie-Mohr, D. and Smith, W. (1999), *Fostering Sustainable Behaviour: An Introduction to Community Based Social Marketing* (Babriola Island, B.C.: New Society).

McKnight, T. (1995), *Oceania: The Geography of Australia, New Zealand, and the Pacific Islands* (New Jersey: Prentice Hall).

Meadowcraft, J. (1998), Co-operative Management Regimes: A Way Forward? in Glasbergen, P. (ed.), *Co-operative Environmental Management Agreements as a Policy Strategy* (Dordrecht, London: Kluwer). pp.21-42.

Meehan, D. (1996), 'The Waste Management Act 1996: The Last Green Bottle', *Irish Planning and Environmental Law Journal*, 3:2, 59-67.

Melosi, M. (2001), *Effluent America: Cities, Industry, Energy and the Environment* (Pittsburg: University of Pittsburg Press).

_____ (2004), *Garbage in the Cities: Refuse, Reform and the Environment* (Pittsburg: University of Pittsburg Press).

Memon, P. and Gleeson, B. (1995), 'Towards a New Planning Paradigm? Reflections on New Zealand's Resource Management Act 1991', *Environment and Planning B: Planning and Design* 22:1, pp.102-124.

Memon, P. and Perkins, H. (2000), 'Environmental Planning and Management: The Broad Context', in Memon, P. and Perkins, H. (eds.), *Environmental Planning and Management in New Zealand* (Palmerston North: Dunmore Press). pp.11-20.

Mercer, C. (1999), 'Reconceptualising State-Society Relations in Tanzania: Are NGOs Making a Difference?', *Area* 31:3, 247-58.

_____ (2003) 'Performing Partnership: Civil society and the Illusions of Good Governance in Tanzania', *Political Geography* 22:7, 741-63.

MfE (1995), *Environment 2010* (Wellington: Mfe - Ministry for the Environment).

_____ (1996), *Notification under the Resource Management Act 1991, Working Paper 6* (Wellington: Mfe - Ministry for the Environment).

_____ (1997a), *National Waste Database* (Wellington: Mfe - Ministry for the Environment).

_____ (1997b), *State of the Environment Report* (Wellington: Mfe - Ministry for the Environment).

_____ (1998a), *Environment 2020 Strategy Stocktake* (Wellington: Mfe - Ministry for the Environment).

_____ (1998b) *National Strategy for Environmental Education – Learning to Care for Our Environment* (Wellington: Mfe - Ministry for the Environment).

_____ (1999) *Your Guide to the Resource Management Act: An Essential Reference for People Affected by or Interested in the Act* (Wellington: Mfe - Ministry for the Environment).

_____ (2000) *Towards a Waste Minimisation Strategy* (Wellington: Mfe - Ministry for the Environment).

_____ (2001) *Our Clean, Green Image: What's it Worth* (Wellington: Mfe - Ministry for the Environment).

_____ (2002) *The New Zealand Waste Strategy: Towards Zero Waste and a Sustainable New Zealand* (Wellington: Mfe - Ministry for the Environment).

_____ (2005) *Waste Management in New Zealand: A Decade of Progress* (Wellington: Mfe - Ministry for the Environment).

_____ (2006a) *Gentle Footprints: Boots 'n' All* (Wellington: Mfe - Ministry for the Environment).

_____ (2006b) *Kiwis take on Environment Message* (Wellington: Mfe - Ministry for the Environment).

Mitchell, F. and Ryan, M. (1997), *Reading the Irish Landscape* (Dublin: Town House).

MOE (2002), *The Environment in Israel 2002* (Jerusalem: MOE - Ministry of the Environment).

Motherway, B., Kelly, M., Faughnan, P. and Tovey, H. (2003), *Trends in Irish Environmental Attitudes Between 1993 and 2002: First Report of National Survey Data* (Dublin: EPA – Environmental Protection Agency).

Murdoch, J. (2004), 'Putting Discourse in its Place: Planning, Sustainability and the Urban Capacity Study, *Area* 36:1, 50-8.

Murphy, P. (1993), *The Garbage Primer: A Handbook for Citizens* (New York: League of Women Voters/Lyons and Burford).

Murray, R. (1999), *Creating Wealth from Waste* (London: Demos).

Murray, J. and Swaffield, S. (1994), 'Myths for Environmental Management: A Review of the Resource Management Act 1991', *New Zealand Geographer* 50:1, 48-52.

Myers, G. (2005), *Disposable Cities: Garbage, Governance and Sustainable Development in Urban Africa* (Aldershot: Ashgate).

Nelson, R. and Winter, S. (1982), *An Evolutionary Theory of Economic Change* (Cambridge, MA: Harvard University Press).

Newell, P. (2000), *Climate for Change: Non-state Actors and the Global Politics of the Greenhouse* (Cambridge: Cambridge University Press).

Nieves, L., Himmelberger, J., Ratick, S. and White, A. (1992), 'Negotiated Compensation for Solid Waste Disposal Facility Siting – an Analysis of the Wisconsin Experience', *Risk Analysis* 12:4, 505-11.

Nissim, I., Shohat, T. and Inbar, Y. (2005), 'From Dumping to Sanitary Landfills – Solid Waste Management in Israel', *Waste Management* 25:3, 323-7.

O' Donovan, O. and Ward, E. (1999), 'Networks of Women's Groups in the Republic of Ireland', in Galligan, Y., Ward, E. and Wilford, R. (eds.), *Contesting Politics: Women in Ireland, North and South* (Boulder, CO: Westview). pp.90-108.

O' Brien, M. (1999), 'Rubbish Values: Reflections on the Political Economy of Waste', *Science and Culture* 8:3, 269-95.

_____ (1999b), 'Rubbish-Power: Towards a Sociology of the Rubbish Society', in Hearn, J. and Roseneil, S. (eds.), *Consuming Cultures: Power and Resistance* (London: Macmillan). pp.262-77.

O'Callaghan-Platt, A. and Davies, A. (2006), *A National Evaluation of the Pay-by-use System* (Wexford: EPA – Environmental Protection Agency).

Ó Gráda, C. (2004), *Ireland's Great Famine: An Overview* (Department of Economics, University College Dublin, Dublin).

O' Leary, E. (2003), 'A Critical Evaluation of Irish Regional Policy', in O'Leary, E. (ed.), *Irish Regional Development: A New Agenda* (Dublin: Liffey Press). pp.15-37.

O' Neill, K. (2000), *Waste Trading Among Rich Nations: Building a New Theory of Environmental Regulation* (Cambridge, MA: MIT Press).

O' Riordan, T. and Church, C. (2001), 'Synthesis and Context', in O' Riordan, T. (ed.), *Globalism, Localism and Identity: Fresh Perspectives on the Transition to Sustainability* (London: Earthscan). pp.3-24.

O' Riordan, T. and Jordan, A. (1996), 'Social Institutions and Climate Change', in O' Riordan, T. and Jager, J. (eds.). *Politics of Climate Change: a European Perspective* (London: Routledge). pp.228-67.

OECD (1998), *Working Party on Pollution Prevention and Control: Considerations for Evaluating Waste Minimisation in OECD Member Countries* (Paris: OECD).
_____ (2004), *Addressing the Economics of Waste* (Paris: OECD).

OECD/Eurostat (2007), *Europe in Figures: Eurostat Year Book 2006-7* (Luxembourg: Eurostat).

Olofsson, J. and Sandow, E. (2003), *Towards a More Sustainable City Planning: A Case Study of Dar es Salaam, Tanzania, Field Report* (Umea: Umea University).

Ong, S. (2001), 'I've Got More Experts Than You – Experts and the Environment Court', *New Zealand Journal of Environmental Law* 5, 261.

Onibukum, A. and Kumuyi, A. (1999), 'Governance and Waste Management in Africa', in Onibokum, A. and Ottawa, A. (eds.), *Managing the Monster: Urban Waste and Governance in Africa* (Canada: IDRC).

Ortiz, J. (2003), 'The Tribal Environment: Solid Waste in Indian Country', *Environmental Management* 31:3, 355-64.

Osbourne, S. (2000), *Public Private Partnerships* (London: Routledge).

OED (1998), *Oxford English Dictionary* (Oxford: Oxford University Press).

Pardilla, J. (1999), National Agenda, *Proceedings of the Second Annual New England Regional Indian Environmental Training Conference, 7 June 1999*.

Parto, S. (2005), *Good Governance and Policy Analysis* (Maastricht: MERIT – Infonomics Research Memorandum Series).

Paterson, M. (1996), *Global Warming and Global Politics* (London: Routledge).

Patterson, J. (1992), *Exploring Maori Values* (Palmerston North: Dunmore Press).

Pawson, E. and Brooking, T. (2002), *Environmental Histories of New Zealand*, (Oxford: Oxford University Press).

PCE (2002), *Creating Our Future: Sustainable Development for New Zealand* (Wellington: PCE – Parliamentary Commissioner for the Environment).

Peart, R. (2004), *Community Guide to the Resource Management Act 1991* (Auckland: Environmental Defence Society).

Pelling, M. (2003), 'Toward a Political Ecology of Urban Environmental Risk', in Zimmerer, K. and Bassett, T. (2003), *Political Ecology: An Integrative Approach to Geography and Environment-Development Studies* (New York: Guildford). pp.73-93.

Pepper, D. (1999), 'Ecological Modernisation or the 'Ideal Model' of Sustainable Development' Questions Prompted at Europe's Periphery', *Environmental Politics* 8:4, 1-34.

Perkins, H. and Thorns, D. (2001), 'A Decade On: Reflections on the Resource Management Act 1991 and the Practice of Urban Planning in New Zealand', *Environment and Planning B: Planning and Design* 28:5, 639-54.

Perreault, T. (2003), 'Changing Places: Transnational Networks, Ethnic Politics and Community Development in the Ecuadorian Amazon', *Political Geography* 22:1, 61-88.

Peters, G. (2000), 'Governance and Comparative Politics', in Pierre, J. (ed.), *Debating Governance: Authority, Steering and Democracy* (Oxford: Oxford University Press). pp.36-53.

Petts, J. (1992), 'Incineration Risk Perceptions and Public Concern: Experiences in the UK Improving Risk Communication', *Waste Management and Research* 10, 169-182.

Pickerill, J. (2000), 'Environmentalists and the Net: Pressure Groups, New Social Movements and New ICTs', in Ward, S. and Gibson, R. (ed.) *Reinvigorating Democracy? British Politics and the Internet* (Aldershot: Ashgate). pp.129-50.

Pierre, J. (2000), 'Conclusions: Governance Beyond State Strength', in Pierre, J. (ed.), *Debating Governance* (Oxford: Oxford University Press). pp.241-6.

Pierre, J. and Peters, G. (2000), *Governance, Politics and the State* (Basingstoke: Macmillan).

Plummer, J. (2002), *Focusing Partnerships: a Sourcebook for Municipal Capacity Building in Public-Private Partnerships* (London: Earthscan).

Porter, G., Welsh Brown, J. and Chasek, P. (2000), *Global Environmental Politics*, (3rd Edition) (Boulder, Co: Westview Press).

Price, J. and Joseph, J. (2000), 'Demand Management a Basis for Waste Policy: A Critical Review of the Applicability of the Waste Hierarchy in Terms of Achieving Sustainable Development', *Sustainable Development* 8:2, 96-105.

Princen, T. and Finger, M. (1994), *Environmental NGOs in World Politics: Linking the Local and the Global* (London: Routledge).

Raco, M. (2003), 'Governmentality, Subject-building and the Discourses and Practices of Devolution in the UK', *Transactions of the Institute of the Institute of British Geographers*, 28:1, 75-95.

Rasmussen, T. (2000), 'State Regulatory Principals and Local Bureaucratic Agents – the Politics of Solid Waste Management', *American Review of Public Administration*, 30:3, 292-306.

Rathi, S. (2006), 'Alternative Approaches to Better Solid Waste Management in Mumbai, India', *Waste Management* 26:10, 1192-200.

Rathje, W. and Cullen, M. (2001), *Rubbish!: The Archaeology of Garbage* (Tucson: University of Arizona Press).

Rhodes, R. (1996), 'The New Governance: Governing Without Government', *Political Studies*, XLIV, 652-67.

_____(1997), *Understanding Governance: Policy Networks, Governance, Reflexivity and Accountability* (Bristol PA: Open University Press).

_____(2000) 'Governance and Public Administration', in Pierre, J. (ed.), *Debating Governance: Authority, Steering and Democracy* (Oxford: Oxford University Press). pp.54-90.

Roome, N. (1998), *Sustainability Strategies for Industry: The Future of Corporate Practice* (Washington, DC: Island Press).

Rosell, E. (1996), 'The Sunland Park Camino Real Partnership: Landfill Politics in a Border Community', *Policy Studies Journal* 24:1, 111-22.

Rosenau, J. (1992), 'Governance, Order, and Change in World Politics', in Rosenau, J. and Czempiel, E. (eds.), *Governance Without Government: Order and Change in World Politics* (Cambridge: Cambridge University Press). pp.1-29.

_____ (1995), 'Governance in the Twenty-first Century', *Global Governance,* 1:1, 13-43.

_____ (2000a), 'Change, Complexity and Governance in Globalizing Space', in Pierre, J. (ed.). *Debating Governance* (Oxford: Oxford University Press). pp.167-200.

_____ (2000b), *Public-Private Partnerships* (Cambridge Ma: MIT Press).

Rose-Redwood, R. (2006), 'Governmentality, Geography, and the Geo-coded World', *Progress in Human Geography,* 30:4, 469-86.

Royte, E. (2005), *Garbage Land: On the Secret Trail of Trash* (New York: Little Brown and Company).

Sabatier, P. (1979), 'Major Sources On Environmental Politics: 1974-1977. The Maturing of a Literature', *Policy Studies Journal* 7:3, 582-604.

Scanlan, J. (2005), *On Garbage* (London: Reaktion Press).

Scannell, Y. (1982), *The Law and Practice Relating to Pollution Control in Ireland,* (London: Graham and Trotman).

_____ (1990), 'Legislation on Toxic Waste Disposal in Ireland', *Papers from the Joint Conference with the Incorporated Law Society of Ireland.*

Schmitter, P. (2001), *What is There to Legitimise in the European Union...and How Might this be Accomplished?* (Vienna: Political Science Series, Institute for Advanced Studies).

Scholefield, G. (1909), *New Zealand in Evolution* (London: Unwin).

Schout, A. and Jordan, A. (2003), *Co-ordinated European Governance: Self-organizing or Centrally Steered* (Norwich: CSERGE Working Paper EDM 03-14).

Schreurs, M. and Economy, E. (1997), *The Internationalization of Environmental Protection* (Cambridge: Cambridge University Press).

Schroeder, R. (1999), Geographies of Environmental Intervention in Africa, *Progress in Human Geography,* 23:3, 359-78.

Scott, W. (2001), *Institutions and Organizations* (London: Sage Publications).

Seadon, J. (2006), 'Integrated Waste Management – Looking Beyond the Solid Waste Horizon', *Waste Management,* 26:12, 1327-36.

Seadon, J. and Stone, L. (2003), The Integration of Waste Management, *Air and Waste Management Association 96th Annual Conference and Exhibition,* San Diego, United States of America, 22-26 June, 2003.

SEPA (2003), *National Waste Management Plan* (Stirling: SEPA - Scottish Environment Protection Agency).

Shields, R. (1991), *Places on the Margin: Alternative Geographies of Modernity* (London: Routledge).

Singleton, S. (2000), 'Co-operation or Capture? The Paradox of Co-management and Community Participation in Natural Resource Management and Environmental Policy-making', *Environmental Politics* 9:2, 1-21.

Smith, A., (1997), *Integrated Pollution Control: Change and Continuity in the Industrial Pollution Policy Network* (Aldershot: Ashgate).

Smouts, M. (1998) 'The Proper Use of Governance in International Relations', *International Social Science Journal,* 50:155, 81-89.

Snary, C. (2004), 'Understanding Risk: The Planning Officers' Perspective', *Urban Studies* 41:1, 33-5.

Snel, M. (1999), 'Integration of the Formal and Informal Sector – Waste Disposal in Hyderabad, India', *Waterlines*, 17:1, 27-9.

Spargaaren, G., Mol, A. and Buttel, F. (2000), 'Introduction: Globalisation, Modernity, and the Environment', in Spargaaren, G., Mol, A. and Buttel, F. (eds.), *Environment and Global Modernity* (Thousand Oaks, CA: Sage). pp.1-15.

Srivastava, P., Kulshreshtha, K., Mohanty, C., Pushpangadan, P. and Singh, A. (2005), 'Stakeholder-based SWOT Analysis for Successful Municipal Solid Waste Management in Lucknow, India', *Waste Management*, 25:5, 531-7.

Star, P. and Lochhead, L. (2002), 'Children of the Burnt Bus: New Zealanders and the Indigenous Remnant, 1880-1930', in Pawson, E. and Brooking, T. (eds.), *Environmental Histories of New Zealand* (Oxford: Oxford University Press). pp.119-35.

Stoker, G. (2000), 'Urban Political Science and the Challenge of Urban Governance', in Pierre, J. (ed.), *Debating Governance* (Oxford: Oxford University Press). pp.91-109.

Stokes, E. (2002), 'Contesting Resources: Māori, Pākehā, and a Tenurial Revolution' in Pawson, E. and Brooking, T. (eds.), *Environmental Histories of New Zealand* (Oxford: Oxford University Press). pp.35-51.

Stone, L. (2002), *Assessment of Waste Minimisation Activities in New Zealand, Centre for Advanced Engineering Resource Stewardship/Waste Minimisation Project, Phase 1 Report* (Christchurch: Centre for Advanced Engineering).

_____ (2003), *Resource Stewardship and Waste Minimisation: Towards a Sustainable New Zealand* (Christchurch: Centre for Advanced Engineering).

Strasser, S. (1999), *Waste and Want: A Social History of Trash* (New York: Metropolitan Books).

Sturman, A. and Spronken-Smith, R. (2001), *The Physical Environment: A New Zealand Perspective* (Oxford: Oxford University Press).

Swyngedouw, E. (2000), 'Authoritarian Governance, Power and the Politics of Rescaling', *Environment and Planning D: Society and Space*, 18:1, 63-76.

Symes, D. (2006), 'Fisheries Governance: A Coming of Age for Fisheries Social Science?', *Fisheries Research* 81:2-3, 113-7.

Tammemagi, H. (1999), *The Waste Crisis: Landfills, Incinerators and the Search for a Sustainable Future* (Oxford: Oxford University Press).

Taylor, G. (2001), *Conserving the Emerald Tiger: The Politics of Environmental Regulation in Ireland* (Galway: Arlen House).

Taylor, R. (2005), 'The Sustainable Households Programme – Education for Taking Practical Steps at Home Towards Sustainability', *Waste Awareness*, Jan-Mar, 14-5.

Te Tari Tatatu (2005), *National Population Statistics* (Wellington: Statistics New Zealand).

Thomas, C. (1999), 'Waste Management and Recycling in Romania: A Case Study of Technology Transfer in an Economy in Transition', *Technovation*, 19:6, 365-71.

Thompson, M. (1979), *Rubbish Theory: The Creation and Destruction of Value* (New York: Oxford University Press).

Tirado, M. (2001), 'The New Indian Wars: Tribes Fight to Restore the Environment', *American Indian Report* 17:7, 12.

Uitermark, J. (2005), 'The Genesis and Evolution of Urban policy: A Configuration of Regulationist and Governmentality Approaches', *Political Geography*, 23:2, 137-63.

UNEP (1989), *Basel Convention on the Control of Transboundary Movements of Hazardous Wastes and Their Disposal* (Nairobi: UNEP).

UNEP/GRID-Arendal (2004), Municipal Solid Waste Composition: for 7 OECD Countries and 7 Asian Cities, *UNEP/GRID-Arendal Maps and Graphics Library*, 2004,<http://maps.grida.no/go/graphic/municipal_solid_waste_composition_for_7_oecd_countries_and_7_asian_cities> [Accessed 14 March 2007]

UNEP/UNSD (2004), *Waste Definitions* (Nairobi: UNEP).

United Nations (2005), *World Population Prospects: The 2004 Revision Analytical Report* (New York: United Nations Department of Economic and Social Affairs: population division).

Upton, S. (1991), *Third Reading Debate on the Resource Management Bill (Debates 516)* (Wellington: House of Representatives).

USHHS (US Department of Health and Human Services) (1998), *Report on the Status of Open Dumps on Indian Lands* (Rockville, Maryland: Indian Health Service).

van Kersbergen, K. and van Waarden, F. (2004), 'Governance as a Bridge Between Disciplines', *European Journal of Political Research* 43:2, 143-71.

Vidanaarachchi, C., Yuen, S. and Pilapitiya, S. (2006), 'Municipal Solid Waste Management in the Southern Province of Sri Lanka: Problems, Issues and Challenges', *Waste Management* 26:8, 920-30.

Viney, M. (2003), *Ireland: A Smithsonian Natural History* (Belfast: Blackstaff Press).

Vogler, J. (2005), 'The European Contribution to Global Environmental Governance', *International Affairs*, 81:4, 835-50.

Walker, L., Cocklin, C. and Le Heron, R. (2000), 'Regulating for Environmental Improvement in the New Zealand Forestry Sector', *Geoforum* 31:3, 281-97.

Walsh, E., Warland, R. and Clayton-Smith, D. (1997) *Don't Burn It Here: Grassroots Challenges to Trash Incinerators* (Pennsylvania: Penn State Press).

Wangwe, S., Semboja, H. and Timbandebage, P. (1998), *Transitional Economic Policy and Policy Options in Tanzania* (Dar es Salaam: Mkuki na Nyota Publishers).

Wapner, P. (1998), 'Reorienting State Sovereignty: Rights and Responsibilities in the Environmental Age', in Liftin, K. (ed.), *The Greening of Sovereignty in World Politics* (Cambridge MA: MIT Press). pp.275-97.

Watts, M. (2003), 'Development and Governmentality', *Singapore Journal of Tropical Geography*, 29:2, 195-216.

WCED (1987), *Our Common Future* (Oxford; Oxford University Press).

Welch, R. (2002), 'Legitimacy of Rural Local Government in the New Governance Environment', *Journal of Rural Studies*, 18:4, pp.443-59.

Wheen, N. (2002), 'A History of New Zealand Environmental Law', in Pawson, E. and Brooking, T. (eds.), *Environmental Histories of New Zealand* (Oxford: Oxford University Press). pp.261-74.

White, L. and du Preez, L. (2005), Measuring the Community Sector's Contribution to New Zealand's Waste Strategy, *Zero Waste Conference*, Kaikoura, 5-8th April 2005.

White, P. (1996), *So What is Integrated Waste Management? Warmer Bulletin: Journal of the World Resource Foundation, No. 49* (Tonbridge, Kent: World Resource Foundation).

Whitehead, M. (2007), *Spaces of Sustainability: Geographical Perspectives on the Sustainable Society* (London: Routledge).

Wicklow County Council (2000), *Wicklow Waste Management Plan 2000-2004* (Wicklow: Wicklow County Council).

Williams, P. (2005), *Waste Treatment and Disposal* (Chichester: Wiley).

Wilson, D. (1977), 'History of Solid Waste Management', in Wilson, D. (ed), *Handbook of Solid Waste Management* (New York: Van Nostrand Reinhold). pp.1-9.

Wilson, E., McDougall, F. and Willmore, J. (2001), 'Euro-trash: Searching Europe for a more Sustainable Approach to Waste Management', *Resources Conservation and Recycling*, 31:4, 327-46.

Wilson, R. (1982), *From Manapouri to Aramoana: The Battle for New Zealand's Environment* (Auckland: Earthworks Press).

Wood, A. (2005), *Demystifying 'Good Governance': An Overview of World Bank Governance Reforms and Conditions* (Dublin: Trocaire).

World Bank (1994), *Governance – The World Bank's Experience, Development in Practice* (Washington: World Bank).

Wynn, G. (2002), 'Destruction Under the Guise of Improvement? The Forest, 1840-1920', in Pawson, E. and Brooking, T. (eds.), *Environmental Histories of New Zealand* (Oxford: Oxford University Press). pp.100-16.

Yearley, S. (1995), 'Dirty Connections: Transnational pollution', in Allen, J. and Hamnett, C. (eds.), *A Shrinking World?* (Oxford: Open University Press).

Yin, R. (1984), *Case Study Research: Design and Methods* (London: Sage).

Young, O. (1997) *Global Governance: Drawing Insights from the Environmental Experience* (Cambridge MA: MIT Press).

_____ (1999), *The Effectiveness of International Environmental Regimes: Causal Connections and Behavioural Mechanisms* (Cambridge MA, MIT Press).

Zeiss, C. (1998), 'Noxious Facilities and Host Community Response: A Causal Framework', *Journal of Environmental Health* 61, 18-28.

Zeiss, C. and Paddon, B. (1992), Management Principles for Negotiating Waste Facility Siting Agreements', *Journal of the Air and Waste Management Association* 42, 1296-304.

Zero Waste Alliance (2004), '*What is Waste?*' (Auckland: Zero Waste Alliance).

Zero Waste New Zealand (2001), *The End of Waste: Zero Waste by 2020* (Auckland: Zero Waste New Zealand Trust).

_____ (2003), *Getting There! The Road to Zero Waste: Strategies for Sustainable Communities* (Auckland: Zero Waste New Zealand Trust).

Zimmerer, K. and Bassett, T. (2003), *Political Ecology: An Integrative Approach to Geography and Environment-Development Studies* (New York: Guildford).

Zurbrügg, C., Drescher, S., Patel, A. and Sharatchandra, H. (2004), 'Decentralised Composting of Urban Waste – an Overview of Community and Private Initiatives in Indian Cities', *Waste Management* 24:7, 2004, 655-62.

Index